DEMOCRACY IN A TECHNOLOGICAL SOCIETY

PHILOSOPHY AND TECHNOLOGY
VOLUME 9

Series Editor: PAUL T. DURBIN

Editorial Board

Albert Borgmann, *Montana*
Mario Bunge, *McGill*
Edmund F. Byrne, *Indiana –
  Purdue at Indianapolis*
Stanley Carpenter, *Georgia Tech*
Robert S. Cohen, *Boston*
Ruth Schwartz Cowan, *SUNY –
  Stony Brook*
Hubert L. Dreyfus, *California –
  Berkeley*
Bernard L. Gendron, *Wisconsin –
  Milwaukee*
Ronald Giere, *Minnesota*
Steven L. Goldman, *Lehigh*
Virginia Held, *CUNY*
Gilbert Hottois, *Université Libre de
  Bruxelles*
Don Ihde, *SUNY – Stony Brook*
Melvin Kranzberg, *Georgia Tech*
Douglas MacLean, *Maryland,
  Baltimore County*

Joseph Margolis, *Temple*
Robert McGinn, *Stanford*
Alex Michalos, *Guelph*
Carl Mitcham, *Pennsylvania State
  University*
Joseph Pitt, *Virginia Polytechnic*
Friedrich Rapp, *Dortmund*
Nicholas Rescher, *Pittsburgh*
Egbert Schuurman, *Technical
  University of Delft*
Kristin Shrader-Frechette, *South
  Florida*
Elisabeth Ströker, *Cologne*
Ladislav Tondl, *Czechoslovak
  Academy of Science*
Marx Wartofsky, *CUNY*
Caroline Whitbeck, *MIT*
Langdon Winner, *RPI*
Walter Ch. Zimmerli, *Bamberg*

The titles published in this series are listed at the end of this volume.

OFFICIAL PUBLICATION OF
THE SOCIETY FOR PHILOSOPHY AND TECHNOLOGY

PHILOSOPHY AND TECHNOLOGY
VOLUME 9

DEMOCRACY IN A TECHNOLOGICAL SOCIETY

Edited by

LANGDON WINNER

Rensselaer Polytechnic Institute, Troy, U.S.A.

KLUWER ACADEMIC PUBLISHERS
DORDRECHT / BOSTON / LONDON

Library of Congress Cataloging in-Publication Data

```
Democracy in a technological society / edited by Langdon Winner.
      p.   cm. -- (Philosophy and technology ; v. 9)
   "Official publication of the Society for Philosophy and
Technology."
   Includes index.
   ISBN 0-7923-1995-8 (hardback : acid-free paper)
   1. Technology and state.  2. Technology--Philosophy.  I. Winner,
Langdon.  II. Society for Philosophy and Technology (U.S.)
III. Series.
T14.D36  1992
306.4'6--dc20                                             92-31873
```

ISBN 0-7923-1995-8

Published by Kluwer Academic Publishers,
P.O. Box 17, 3300 AA Dordrecht, The Netherlands.

Kluwer Academic Publishers incorporates
the publishing programmes of
D. Reidel, Martinus Nijhoff, Dr W. Junk and MTP Press.

Sold and distributed in the U.S.A. and Canada
by Kluwer Academic Publishers,
101 Philip Drive, Norwell, MA 02061, U.S.A.

In all other countries, sold and distributed
by Kluwer Academic Publishers Group,
P.O. Box 322, 3300 AH Dordrecht, The Netherlands.

Printed on acid-free paper

All Rights Reserved
© 1992 Kluwer Academic Publishers
No part of the material protected by this copyright notice may be reproduced or
utilized in any form or by any means, electronic or mechanical,
including photocopying, recording or by any information storage and
retrieval system, without written permission from the copyright owner.

Printed in the Netherlands

TABLE OF CONTENTS

PREFACE, Paul T. Durbin — vii

INTRODUCTION, Langdon Winner — 1

PART I
THE NATURE OF THE PROBLEM

MARX W. WARTOFSKY / Technology, Power, and Truth: Political and Epistemological Reflections on the Fourth Revolution — 15
JACQUES ELLUL / Technology and Democracy — 35
ROBERT L. FROST / Mechanical Dreams: Democracy and Technological Discourse in Twentieth-Century France — 51

PART II
SOME PROPOSED SOLUTIONS

PAUL T. DURBIN / Marxism and the Democratic Control of Technology — 81
LARRY A. HICKMAN / Populism and the Cult of the Expert — 91
JOHN FIELDER / Autonomous Technology, Democracy, and the Nimbys — 105
KRISTIN SHRADER-FRECHETTE / Technology, Bayesian Policymaking, and Democratic Process — 123
RICHARD E. SCLOVE / The Nuts and Bolts of Democracy: Democratic Theory and Technological Design — 139

PART III
HISTORICAL AND CULTURAL REFLECTIONS

STANLEY R. CARPENTER / Instrumentalists and Expressivists: Ambiguous Links Between Technology and Democracy — 161
VITALY GOROKHOV / Politics, Progress, and Engineering: Technical Professionals in Russia — 175
TOM ROCKMORE / Heidegger on Technology and Democracy — 187
ALBERT BORGMANN / The Moral Assessment of Technology — 207
JOSEPH MARGOLIS / Political Morality under Radical Conditions — 215

NAME INDEX — 235

PREFACE

This ninth volume is one of the most ambitious in the Philosophy and Technology series. Edited by technopolitical philosopher Langdon Winner, it assembles an impressive collection of philosophers and political theorists to discuss one of the most important topics of the end of the twentieth century – the bearing of technology, in all its ramifications, on the practice of democratic politics in the developed world. When set beside the previous volume in the series – *Europe, America, and Technology* – the two together open a philosophical dialogue of great significance about the ways technology challenges democracy at its very roots. Some philosophers think the attack is fatal. Others are optimistic that democratic means can be discovered, or invented, for the control of technology. Still others object to an optimism-versus-pessimism formulation of the issue. But all agree that the issue is highly significant, one that demands serious philosophical inquiry. The Society for Philosophy and Technology was fortunate in being able to draw this group of writers to Bordeaux, France, in 1989, along with a large number of others whose contributions to the debate could not be included here. It is equally fortunate to have chosen Langdon Winner as president when the time came to select the best of the papers to fashion this volume.

University of Delaware PAUL T. DURBIN

LANGDON WINNER

INTRODUCTION

What is the relationship between democracy and technology? And what should that relationship be? The essays gathered here explore these questions, drawing upon a wide range of philosophical, historical, and sociological points of view. The authors arrive at no firm consensus on the matter. Indeed, there is considerable disagreement among them about which ideas, arguments, and theoretical positions offer the most fruitful results. Their common ground is a desire to rethink the horizons of democratic politics in an age increasingly shaped by the power of technologically embodied institutions and practices.

Earlier drafts of these essays were presented at a conference on Technology and Democracy sponsored by the Society for Philosophy and Technology at the University of Bordeaux in June 1989. It was the bicentennial of the French Revolution, a time to celebrate the event and ponder its consequences for France and the rest of the world as well. But in ways that soon revealed eerie parallels between rebellions across two centuries, political upheavals that year ruptured the boundaries of festival. As the conferees gathered in Bordeaux and as lavish ceremonies for Bastille Day were being orchestrated in Paris, recent news from Tiananmen Square in Beijing, the massive use of tanks and military troops to crush the pro-democracy movement, was painfully fresh in peoples' minds. Democracy, expressed as the demand for freedom of speech, press, and public assembly, had been brutally denied to the people of the world's most populous nation state.

Autumn of that year brought similar political crises in several other countries. With astonishing speed pro-democracy forces toppled Communist regimes in East Germany and several other Eastern European countries. As the Berlin Wall was suddenly reduced to rubble, the word "democracy" was on everyone's lips, the name of an imagined promised land that lay beyond decades of totalitarian repression. Of course, the democratic ferment did not end there. Within two years the bulwark of European Communism, the Soviet Union, had itself collapsed as a result of internal pressures that took democratization as a rallying cause.

Each of the eruptions brought extravagant hopes of political, social, and economic renewal. While it is too early to tell what the fate of the post-Communist rebuilding will be, the hopes that surrounded pro-democracy movements in their initial stages have been overshadowed by multiple crises of economic collapse, political disorder, environmental devastation, ethnic strife, and armed aggression. The possibility of achieving peaceful, prosperous, well-governed societies turned out to be far more difficult than the rebels first expected. As the experience of the French Revolution had made clear, beyond the destruction of the old regime lie many pitfalls.

For those who study the prospects for democracy in the late twentieth century, questions about technology and politics are never far from the surface. Thus, many commentaries on the collapse of Communism focused on the powerful solvent that electronic information systems had poured over the entrenched authoritarian bureaucracies of the U.S.S.R. Within the boundary-busting age of small portable computers, copying machines, and satellite television, it was no longer possible to control information in a highly centralized, Stalinist manner. At first it seemed that similarly disruptive forces would prove decisive in China. During the peak of the Tiananmen Square demonstrations, support from around the world poured in via fax machine, a device whose use had quickly spread during the late 1980s. International satellite television both covered and encouraged the demonstrations up to the point where Chinese government officials appeared (literally before the cameras) to shut things down, a mass media prelude to the bloody suppression to follow. Euphoria among Western observers about the liberating role of electronic communications was later dampened by reports that Chinese authorities had been able to round up leaders of the demonstration by tracing the destinations of fax transmissions.

The discovery of strong tensions between technological innovation and the promise of democracy is by no means new. Modern political structures and practices arose hand in hand with the power of industrial technology. Democratic revolution and industrial revolution were twins of a sort, but ones that from the very beginning lived with troubled relations. For many observers two centuries ago, the democratic principle of equality seemed to find a perfect match in the ability of industrialism to expand material production. Goods once available only to an aristocratic few could now be made for the great mass of people. Improvements in technology promised to eliminate barriers to the achievement of universally shared wealth. This accomplishment, many believed, would make democratic self-government more stable because widely distributed, industrially produced riches would remove the sources of class conflict that had vexed democracy in antiquity.

From its earliest days, however, the promise of democratic

industrialism was haunted by evidence of chronic inequality. As nineteenth-century workers' movements and socialist writers pointed out, the problem was not merely that the fruits of the machine were unevenly distributed, but also that the entire material and institutional system of the new society was organized in ways that favored the interests of some but not others. This fact, it became clear, was of political as well as economic significance, for power within the state was obviously linked to concentrations of industrial wealth. The notion that scientific and technical advance would be uniformly beneficial for the populace as a whole and a reliable aid to self-government soon became the target of sustained attack, expressed in philosophical critiques, utopian speculation, mass political movements, and attempts at political reform. In ways still very much at issue in our time, controversies about what egalitarian principles require for the organization and control of industrial production have long been the centerpiece in debates about democracy and technology. The fact that various Communist and socialist alternatives have now fallen into disfavor has not eliminated the fundamental problems that gave rise to them.

Another long-standing source of strain between modern technological development and the practice of democracy stems from the erosion of citizen participation in decision making when confronted with the evolving power of scientific and technical elites. In its classic form, this became the question of technocracy, the possibility explored by Thorstein Veblen and others that the complexities of an advanced technological society would inevitably bring it under the governance of a social class of well-trained experts. Thinkers who have described this as a problem rather than an opportunity have worried that modes of traditional understanding used by ordinary people would simply be overwhelmed by the rapidly proliferating varieties of scientific and technical knowledge needed to operate modern sociotechnical systems. What would become of a democracy whose citizens no longer understood the fundamentals upon which key decisions are based?

Fears about the rise of a unified, self-conscious technocratic elite have subsided in recent times. Replacing these concerns are questions about the normative and practical foundations of attempts to democratize technological choices. Is it philosophically warranted to conclude that in an age of pervasive, world-shaping technological development, citizens have a right to be directly involved in technological choices that affect them? If one answers "yes" to that question, then means must be found to realize that right in practice. Several of the essays in this volume take up this issue, offering arguments to justify the contributions of populist social movements and to support other varieties of citizen participation in technology policy making and technological design. The writers disagree, however, about whether or not existing institutional forms are adequate for the job of expressing democratic sentiments. Will it be enough to muddle through with piecemeal remedies, or is a more fundamental restructuring required?

A closely related set of debates concerns the politics of technological discourse. In areas of environmental policy, industrial planning, health care, the evaluation of hazards and "risks," military affairs, many of the key disputes hinge upon the categories of analysis and evaluation employed as decisions evolve. What appear to be eminently "rational" and politically "neutral" forms of philosophical and technical analysis can mask biases profoundly unfriendly to democratic principles. Several of the papers in this volume examine concepts commonly used in decision making about technological applications, asking which methods are compatible with norms of equality, freedom, and participation. A number of writers here raise questions about the kinds of moral and political speech needed in a society in which democracy and technological development must somehow coexist.

Another pivotal issue has to do with the general effects new technologies have on the texture of civic life and their consequences for genuine self-government. In many of today's nominally democratic societies, the extent of citizen activity has been stripped to a bare minimum. Citizenship involves only paying one's taxes, obeying the law, and serving in the military, if such is required. Voting has become an option that fewer and fewer find attractive; less than fifty percent of the American electorate has bothered to go to the polls in recent elections. Even those with strong grievances against public officials often feel that voting "would only encourage them" and that, consequently, the best strategy is to avoid the ballot box altogether. Tendencies of this kind cast a shadow over hopes expressed by many twentieth-century political theorists that democracy might be expanded beyond the boundaries of modern republicanism, with its emphasis upon elections and representative government, to develop more direct forms of citizen participation. If politics in its minimal definition is of little concern to people, what real chance is there to broaden the horizons?

Clearly, the influence of technology in its various forms is not the only factor helping to undermine the vitality of contemporary citizenship. A number of deeply rooted problems in the operations of governments, political parties, and interest groups have contributed to a widely felt mood of alienation. But technologies play a distinctive role in this evolving predicament because they help define social membership in powerful ways. To an increasing extent, people see themselves not as citizens, not as public persons in any meaningful sense, but as human components in ongoing processes of technology-centered production and consumption. In terms their leaders use and that most people unthinkingly accept, they are "wage earners" or "consumers" or "taxpayers." What matters is "the economy" and whether or not its prospects at any given time look good or bad. Understood in this way, the health of the polity and the well-being of its members are measured by "leading economic indicators" and the "gross domestic product." The happiness of citizens is to be found in material gratification alone. Missing from the self-identity of most adults in advanced industrial societies is any desire for involvement in public life or any sense that such activity might carry intrinsic rewards.

In fact, for a great majority of the population in many nation states, politics has become a minor subdivision of a much more captivating institution—television. In ways that now seem all but impossible to undo, the pulse of television entertainment has insinuated itself as the dominant style of public expression. Following the lead of MTV, video advertising and infotainment news, political spokesmen now address the public in brief, hard hitting "sound bites." The prevailing assumption is that any idea longer than a sentence is too boring for today's television viewing public to endure. No longer is it necessary to sketch the historical context for a factual statement or to give a reasoned argument for a position. Television viwers are far too impatient to abide those discussions; equipped with hand-held remote control switches, they soon switch to another of several dozen channels available on their TV screens. In this setting the content of public opinion and of electoral communication has become a carefully managed contest of video symbols and increasingly hollow slogans. Indeed, as candidates for public office are finding out, the most effective stance may be simply to project a pleasant image and avoid talking about specific issues at all.

The technically-aided flight from the responsibilities of democracy comes at a time in which the scope and effectiveness of technology yield powers unprecedented in human history. To a greater or lesser degree, all nation states are now competing supplicants to a new ruling force in the world, the global networks of production and communication controlled by transnational business firms whose decisions determine how capital is invested, which research and development is sponsored, which jobs are created, and where production facilities will be opened or closed, and, in effect, which forms of private and public life the future will contain. If our reflections about the future of democracy are to be anything other than impotent fantasies, we must somehow confront this organizing force and its tendency to destroy the ability of political society to chart its own destiny.

In the opening essay here, Marx Wartofsky examines the power of contemporary technology in this light, naming it a "fourth revolution," one that builds upon but radically extends technological and cultural revolutions that occurred in earlier times. He describes the fourth revolution as the "politicization of technology, the concentration of power to make technological policy decisions, and to wield these vast technological forces on a national, international and transnational scale." These developments pose serious problems about the kinds of knowledge required to make rational decisions in the application of technological power. Equally vexing are problems for our understandings of political authority, the making of decisions that are normatively justified. First presented as a presidential address for the Society for Philosophy and Technology at the Bordeaux conference, Wartofsky's paper asks, "How is this to become a democratic revolution?"

The essay that follows was written by one of the twentieth century's most prominent critics of technological society, Jacques Ellul. A lifelong resident and former mayor of the city of Bordeaux, Ellul took the occasion of the Bordeaux conference to offer some candid reflections on challenges that face contemporary society. Arguing that "our Western political institutions are no longer in any sense democratic," he probes the developments that have neutralized modern democracy—the rise of an entrenched political class, the suffocation of the public voice through excessive administration, and reduction of all choices to matters of technique. His purpose is not to bemoan such problems, but to identify possibilities for change. In the essay Ellul sketches a number of proposals—developing citizen competence, restoring classical virtues of self-government, equalizing worldwide living standards, and exploring utopian visions of social transformation—that satisfy the demands of his characteristically uncompromising critique.

The following chapter gives a fascinating historical account of the predicaments with which Ellul and other contemporary French writers have struggled in the domain of social theory. Robert L. Frost begins by comparing two distinctly different symbols of French technological prowess—the Eiffel Tower and the plutonium breeder reactor at Creys-Malville—showing how they symbolize and help constitute the values of emerging political elites. Drawing upon recent studies in the history and sociology of technology, he describes the central conflicts and coalitions involved in the creation of France's modernist technological culture.

There is a striking contrast between the uses of history found in the first two essays and Frost's social constructivist approach. Wartofsky and Ellul look to the past as a setting within which to locate reasoned judgments about how things stand today and what urgently needs to be done. Frost offers highly illuminating interpretations and explanations, but avoids passing judgment on historical developments or their meaning for contemporary choices. Some of the difference can be attributed to disciplinary focus; Wartofsky and Ellul are philosophers, Frost a historian. But the desire to achieve distance from moral and political commitment seems characteristic of constructivist scholarship about science and technology, a reflection of the relativism that underlies its methodology. It remains to be seen what substantive positions on the politics of technology will emerge from constructivist studies.

In one way or another, the next five essays investigate theoretical and practical dimensions of citizen action aimed at influencing the course of technological change. Paul T. Durbin affirms the social meliorist approach set forth in the philosophies of John Dewey and George Herbert Mead, rejecting calls for revolutionary solutions to ills that arise within today's high-technology society. He argues that while Marxist approaches hold out the promise of sweeping social renewal, they tend to overlook maladies involved in attempts to build a classless society, e.g., the suppression of civil liberties. In Durbin's view the adoption of a progressive-liberal strategy need not confine a democratic citizenry to dealing with narrowly defined technosocial issues one at a time. Critical reflection on twentieth-century experience with technology suggests the possibility of creating a worldwide New Progressive Movement with an extremely broad agenda of reform.

Larry A. Hickman's "Populism and the Cult of the Expert" re-examines the ongoing debate about the competence of citizens in technological decision making. This issue flared anew in the late 1960s with the openly technocratic proposals of Emmanuel Mesthene in a report for the Harvard Program on Technology and Society, answered by John McDermott's slam-bang attack in *The New York Review of Books*. Hickman sketches the gap between the two positions, noting that the controversy was ultimately fruitless because neither side offered a coherent vision of technology and democracy. He recommends a more constructive avenue, one that extends Dewey's ideas on the "problems of the public" to chart a truly populist approach to debate and planning.

In what specific ways might contemporary political movements begin to realize the democratic hopes that Dewey, Durbin, and Hickman espouse? John Fielder finds that groups organized around the rallying cry, "Not in My Backyard," the "Nimbys" as they are commonly called, have been able to transcend the kinds of narrow self-interest their name implies and to push public engagement with technological choices to new levels. Seizing issues that involve the siting of large-scale, potentially hazardous technological facilities, such groups have gradually changed from Nimbys to Niabys—"Not in Anybody's Backyard." Along the way, these citizens' movements have helped create new methods and institutions that bring important decisions closer to the reach of ordinary people and local communities. Taken as a whole, Fielder argues, these efforts offer a promising response to "the domination of life by autonomous technology and the large organizations associated with it."

Kristin Shrader-Frechette insists that as popular movements tackle environmental and social problems associated with technology, they must become skillful in making detailed analyses and arguments to counter those of their opponents. She holds that if one is to be effective as an activist, one must enter into a dialogue with engineers, scientists, economists, and government policy makers, learning the categories they employ while not allowing oneself to be taken to the cleaners. Her essay here discusses one Bayesian decision rule frequently used by decision makers in the evaluation of technological risks, the criterion that it is rational to choose the action that produces the best expected value or utility. Weighing philosophical arguments used to defend this criterion, Shrader-Frechette concludes that its justifications are not persuasive. From the standpoint of fundamental democratic principles, the Bayesian rule at issue is deficient and should be rejected in favor of egalitarian norms and open democratic processes in assessing risks.

Richard E. Sclove takes a much different route to public encounters with technological choice. He finds virtually all of the familiar approaches to the politics of technology superficial because they fail to reach the heart of the matter—the design of technology as it affects personal autonomy and the shape of political society. Developing ideas of citizen and community rooted in the philosophies of Jean-Jacques Rousseau and Immanuel Kant, Sclove concludes, "Just as it is democratically essential that citizens be empowered to help shape formal legislative or electoral agendas, it is likewise essential that there be extensive opportunities for citizens to participate in technological design processes." As an aid to thinking about the possibilities here, he offers a system of "rationally-contestable design criteria" that reflect what the quest for a democratic technology and architecture might seek to accomplish in built form.

But why is it that modern societies have so often designed technologies ill-suited to human freedom and the maintenance of a healthy biosphere? Why have we not been able to think more clearly and act more decisively about these matters? Stanley R. Carpenter finds the origins of the problem in a conflict between two strands of moral discourse that have strongly shaped modern thought. One strand, traceable to Enlightenment theories and their defense of political atomism, regards technology in purely instrumental terms. A second tradition, based in Romanticism, sees technology as a mode of expression, a self-defining way of being in the world. Carpenter holds that these ways of thinking exist as an unresolved tension "in our bones," one that generates recurring confusion and ill-fated technological projects. He calls for a novel synthesis of instrumental and expressive modes of understanding and suggests how it might begin.

A new beginning of a painful sort is the topic of the essay on Russian engineering by Vitaly Gorokhov. He notes that in its initial stages the profession developed along lines common to those of engineering in Europe and America. In the years immediately following the October Revolution of 1917, it appeared that such growth might even be enhanced. The Soviet government looked to educated engineers to play an important role in reshaping the economy. But the honeymoon ended with Joseph Stalin's rise to power. Stalin believed that the future of the revolution rested exclusively in the hands of the workers and demanded that suspected bastions of bourgeois privilege be rooted out. Gorokhov vividly describes the calamitous consequences of this policy for Russian engineers in the 1930s and afterwards. Facing a new era in which relationships between technical professionals and the state will have to be reorganized, Gorokhov asks that we recall the philosophical concerns of some of the founders of Russian engineering.

The dark shadow that totalitarian politics has cast over philosophical attempts to understand technology in the twentieth century is also the subject of Tom Rockmore's essay. He asks us to reconsider the thinking of Martin Heidegger, taking into account the fact that "this important German thinker was deeply and irrevocably committed to one of the most antidemocratic movements of this or any other historical moment." Was Heidegger's Nazism an incidental feature of his work, a minor tactical error within the philosopher's otherwise formidable critique of the foundations of modern thought? Or were his hopes for Adolf Hitler and the Nazi Party reflections of a serious defect in his project, one that should cause us to doubt his seemingly profound claims about the essence of technology and the destiny of the world? Rockmore ponders the issue in light of Heidegger's philosophy of Being. He argues that Heidegger's attempt to offer an ontological account of technology is deeply flawed by its abstraction from actual cultural practices, a quality directly linked to Heidegger's embrace of Nazism and contempt for democracy in any form.

The last two essays in the book focus upon the question of morality as regards today's troubled encounters between technology and politics. Albert Borgmann reviews some prominent approaches to assessing technology and concludes that they avoid what should be our central concern. Standards of judgment employed in liberal and environmental assessments may dispute specific technical applications, but they uphold the notion that technology is neutral and instrumental. In Borgmann's view, the conventional approach "leaves the intrinsic moral quality of technological devices unexamined and implicitly sanctions it." He goes on to propose an alternative way of evaluating technological possibilities, quotidian moral assessment, one that takes into account the concrete, multiple ways that technology affects everyday life.

The concluding essay by Joseph Margolis, "Political Morality under Radical Conditions," takes an ambitious route to the question of morality, discussing the intellectual strategies of premodern, modern, and postmodern moral philosophers. The continuing search for moral laws or canons that stand outside specific cultural practices has been, in his view, an entirely fruitless quest. "The artifactual nature of cultural existence precludes any finding of universal law or rule or principle of prudence." While Margolis finds the proposals of Karl Marx and other modernists inadequate for this reason, he does not celebrate postmodernism as a refreshing release from older errors. The postmodern theories of Michel Foucault and Richard Rorty, he maintains, are thoroughly aimless in a moral and political sense. As a starting point for further moral reflection, Margolis asks that we view history as "an order of construction, not of discovery."

The diverse approaches and conclusions of this group of writers converge on perhaps only one point. The comfortable, apparently reasonable vision of progress that once guided political thinking about technological projects and policies has now collapsed. That vision had at its core the faith that conditions of human living inevitably improve, primarily through advances in science and technology and their various economic applications. This conviction arose strongly in the industrial societies of the nineteenth century, and was later adopted as a secular, ecumenical faith around the globe. Those who embraced this idea saw themselves propelled onward and upward by a process larger than themselves, a beneficent historical dynamism whose destiny was to create a better society, perhaps even an earthly paradise.

For serious thinkers today and for growing numbers of politically active persons as well, the progressivist view of history seems neither true nor appealing. The dubious metaphysical underpinnings of the faith are now embarrassingly obvious. Mounting signs of the social and environmental problems that "progress" generates have eroded the conviction that there is a necessary link between technological advance and the well-being of the Earth and its people. Insofar as the aspirations of democratic societies still hinge upon outworn progressivist mythology, those aspirations are in deep trouble. By now the need to formulate new understandings and practical alternatives should be painfully obvious to everyone. The voices assembled here offer a sense of the possibilities for both theory and practice.

ACKNOWLEDGMENTS

Preparations for this book were aided by a year of research leave at the Center for Technology and Culture at the University of Oslo, Oslo, Norway. I gratefully acknowledge the financial support of the Norwegian Research Council for Science and Humanities. The comments and encouragement of Francis Sejersted, David Levinger, Elisabeth Gulbrandsen, Ingunn Moser, and Torben Hviid Nielsen were especially helpful in preparing the manuscript. Larry A. Hickman, co-chair of the program for the Bordeaux conference, offered many valuable suggestions at an early stage in the editing. My special thanks also to Kathrine Høegh-Omdal, Gro Hanne Aas, and Brita Brenna for their intellectual and logistical assistance. Similar assistance was provided at a later stage in the production of the book by Mary Imperatore and Gail Ross of the Philosophy Department at the University of Delaware.

Finally, I wish to express my gratitude to Daniel Cérézuelle for his tireless, effective, and thoroughly gracious efforts as the organizer of the SPT conference on Technology and Democracy in Bordeaux.

Rensselaer Polytechnic Institute

PART I

THE NATURE OF THE PROBLEM

MARX W. WARTOFSKY

TECHNOLOGY, POWER, AND TRUTH: POLITICAL AND EPISTEMOLOGICAL REFLECTIONS ON THE FOURTH REVOLUTION

PREFATORY REMARKS ABOUT THE FOURTH REVOLUTION

Talk about revolutions is easy. Revolutions are hard. On this occasion I have chosen to do the easy thing: to write about a revolution that is in the making, one that I shall call "the fourth revolution." This obviously suggests that there were three others which preceded it, and in due course I will indicate which they were.

This fourth revolution grows out of the confluence of two major world developments in the present: *first*, the proliferation of a global and systematically-linked high technology, which is rapidly becoming a dominant motor force in economic, political, military and social contexts, on a scale and at a pace hitherto unimagined; and *second*, the spread of a world movement of revolutionary social, political and economic change, largely inspired by democratic and national liberation movements, rooted historically in the American and French popular revolutions of the late eighteenth century, in the Russian socialist revolution in the early years of the twentieth century, and in the breakdown of colonialist imperialisms in the mid and late twentieth century. This vast social sea change, however, is also marked by the resurgence of chauvinist nationalisms, religious fundamentalisms, and by the most retrograde expressions of ethnic and racial hatreds, some of which had remained repressed and unresolved for long periods of time.

It is in this coming together of technological development and political change that the prospects and the imperatives of this fourth revolution appear. Three questions emerge, which I will try to address in this paper:

1) The question of the *power* of technology. This divides into two questions: one concerning the technical capacities or powers which the new forms of high technology make available; the other concerning the political, economic, social, or military power which the deployment of this new technology affords.

2) The question of how this power is to be wielded, i.e., what *knowledge* is required to make rational decisions in the exercise of this technological power.

3) The question of *who* has the ability and the authority to exercise

this power, or who *ought* to exercise it.

To summarize my thesis briefly at the outset: Technological development, especially of high technology—has become a question of national and international—or transnational—policy and decision. It has become a central if not yet the dominant question in national and global politics, affecting the basic issues of *power*: political, economic, military and social power—thereby directly affecting the lives and future of people on a mass scale. Control and direction of this technology has therefore become a matter of much greater complexity and scope than it has ever been in the past—one may say that it has become a qualitatively different question than it has been heretofore. These questions concerning *power, knowledge* and *authority* therefore became crucial political question, as well as a technological, or scientific or economic one. Moreover, given the complexities and, as we shall see, the contradictions which emerge in the exercise of this power, a very sharp question arises as to whether rational choices are at all possible, or as to what the conditions are for such rational choice, or of who it is that is capable of making such choices. In short, the epistemological question of what knowledge is, or of what the norms of rationality and of truth are, in the contexts of policy decisions about the development or deployment of this technology, or of the limitations or constraints to be put on its use or its further development—this essentially epistemological question becomes a practical one as well. What knowledge is needed to make such decisions rationally? What decision procedures are adequate to such a task? Who is in a position to acquire this knowledge, and to effect these decisions?

These are basic questions about knowledge and power, questions concerning what we may call political epistemology, of a sort that Francis Bacon had already raised in the transition phase between the first and second revolutions, and which Norbert Wiener had formulated concerning the transition from the second to the third revolutions. They arise again with special urgency at the beginnings of the fourth revolution.

At present, the power—or seeming power—to make such policy decisions about the development and deployment of technology rests with hierarchies or elites in government or in the military or in corporations—in the west, in what has been called the "military-industrial complex"—a term introduced, let it be remembered, by U.S. Republican President Eisenhower, in the last days of his tenure, as a term expressing his concern about the future—in Japan, in the tight-knit and centralized structure of the state, industry and finance; in the (former) Soviet Union or in China, in party, government and military bodies, equally or more hierarchial yet. Ostensibly, the technical knowledge required for policy decisions rests with the experts—scientific-technical personnel who are either part of such hierarchies or are consultant to them. The public is as poorly represented in this decision process as it is more generally in matters of government of corporate or military policy—that is, by whatever minimal degree of representation is afforded by the various political structures, but largely only passively, and as a political mass which has to be persuaded, or cajoled, or neutralized or coerced periodically into accepting, or remaining indifferent to such decision making. Policy-making is generally well insulated by many layers of bureaucratic structure from democratic involvement. Popular democratic responses to technological policy decisions therefore surface primarily in *independent* mass movements—anti-nuclear, environmental, etc.—or on particular issues in political campaigns—on SDI (Star Wars) or acid rain or nuclear waste disposal—or on trade-union issues—on automation, or industrialization policy, etc. Ordinarily, these are one-issue movements. Sometimes however they become articulated political movements which develop a continuity, structure and breadth which joins them to the major world-wide movements for democracy and social change. (In European politics, the Greens represent one such popular long-term response to the hierarchical control of technological decision making.)

The issue is this: since high technology so decisively affects the lives and livelihood and future of people, nationally and globally, and since policy decisions are now on such a general political scale as to become largely national and international governmental decisions, *can* the democratic imperative of self-government by the people be effectively carried into the arena of such technological decision making?

How are those who ultimately produce the technologies, and those who are affected by them in their daily lives, to achieve democratic control of this technological power? How is this question to be addressed concretely and analytically? And if, as I will argue later, the fourth revolution is essentially just this *politicization of technology*, this concentration of the power to make technological policy decisions, and to wield these vast technological forces on a national and international or transnational scale—how is this to become a democratic revolution? Or, more pessimistically, is there an unresolvable contradiction between the political imperatives of democratization and the (allegedly) immanent imperatives of technological development *as such*? (The phrase "as such" connotes some objective, internally driven forces of development which breed just such hierarchical and authoritarian decision-and-control-processes as their objective structural correlate.)

I will propose later, in the conclusion of this paper, that this fourth revolution is fraught with great dangers, that it in fact embodies sharp contradictory or conflicting tendencies, and that its prospects as a *democratic* revolution depend on some hard choices and difficult tasks—mainly on two: (1) the radical democratization of power in society; (2) the education, in a major way, of the scientific and technical understanding of the public to the extent that some forms of democratic participation in scientific-technical policy-making becomes feasible and useful, and not simply an empty populist piety.

I want to put these preliminary considerations about the fourth revolution into historical or genetic perspective, which requires dealing first with the three revolutions that preceded it. For reasons that will become clearer, I hesitate to call these technological revolutions, for fear of suggesting too narrow a comprehension of them. I would rather call them cultural revolutions. But this term is so laden with the horrors of the Chinese "Cultural Revolution," on the one hand, and with the haze which surrounds the word "culture" for another, that I will proceed with "technological," suitably characterized, as my adjective of choice. I begin, then, with an account of technology appropriate to the notion of a first or originary technological revolution. I then go on with a characterization of the second and third revolutions, in order to set the stage for the fourth. Then I will proceed to consider the concept of technological *power*—both as the power *of* technology and power *over* this technology, i.e., the power to effectively use and control it. Finally, I want to show how this fourth revolution, like the others, is not only shaped by human choices, by the human forces of hand and mind, but also reflexively changes the very character of human cognition and action, and operates as a means by which the very nature of human knowledge and self-understanding is transformed—that is, by which human nature changes itself historically. This reflexivity of technology has, as I hope to show, consequences for the epistemological conception of truth and knowledge. After these reflections on *power* and *truth* I will return to the question of the fourth revolution, now considered in the perspective of the foregoing discussion.

TECHNOLOGICAL REVOLUTIONS

Human beings are both creators and creatures of technology. If, by technology, we mean the means and the methods which human beings have invented, or discovered, for effecting their deliberately chosen ends, or for satisfying their needs and wants; and if, by technological change or development we mean both the introduction of new means to achieve either old or new ends, or the improvement of existing means in terms of their effectiveness or efficiency—if, in short, we mean by technology both the hardware and the software—the tools and ways of using them—of effective human action, then we may mark off several major stages of the development of such technology. In the course of the creation and use of these means and methods, human beings have recreated or transformed themselves. The history of this transformation is what we may broadly designate as cultural history. There is a conceptual, and not just a terminological issue to be resolved here, and I can do no more than broach it, since it requires a fuller discussion than I can give it here. It concerns what may be recognized as a distinction between the history of technology in some narrower acceptation—e.g., as a history of the development of the material means of production (whether agrarian, industrial, or domestic), or of computation, or of warfare, or of transportation, etc., or some history of the "mechanical" arts and sciences (or of the electronic ones, for that matter)—and the history of societies, or of ideas, or of the mind, or of culture, regarded as *Geisteswissenschaften*. I use the term *cultural history* inclusively, as a way of connoting both of these emphases, in order to preserve the sense of their relationship—of head and hand, or of *episteme* and *techne*, perhaps in that Hegelian sense in which *Bildung* is as much a matter of cultural and cognitive development as it is of self-constitution by means of social practices and interactions. In the sense that such cultural history is a directed or progressive history, i.e., one in which later stages do not simply succeed earlier ones, but in some sense grow out of them, or elaborate or realize possibilities created by the earlier stages, we may characterize this history as cultural evolution.

For the purposes of this paper, I will take technological evolution to be only an aspect, and not the whole of this cultural evolution. Although it is tempting to interpret technology so broadly as to include the widest variety of means and methods of human cultural activity, I will concentrate here on only certain of these: *first*, on those artifacts made and used for the production (or reproduction) of material and social life; and also the artifacts made and used for the production of other artifacts. These will count as what I will call *tools* (or *means* or *instruments* more generally); and *second*, also the ways, processes, rules of art, and practices which describe the *methods* by which such artifacts are made and used. The generic term *techne* I will take to connote the skill, art, technique or craft that constitutes a technical practice, i.e., an activity in accordance with, or exemplifying a *method*, or *rules of practice*.

Thus, we have three elements of a technology: those artifacts made and used as *tools*; the *methods* or rules of practice of producing and reproducing such artifacts, and of using them; and the *technai* or skills, i.e., the actual activities or practices which are carried out in accordance with such methods, or which embody them.

Tools or instruments need not be "material" artifacts, e.g., physical objects shaped for a certain use. They may be geometrical constructions—e.g., the circle or the right angle, or the 3:4:5 right triangle developed by the ancient Egyptians for constructing the right angle, which was the forerunner of the Pythagorean Theorem. In short, "tools" may be mathematical computational devices or record-keeping devices, and there may be methods or rules of art for such constructions, and the appropriate skills or practices which operate in accordance with these methods, with respect to mathematical formulas.

Let me now briefly sketch the first three technological revolutions, or broadly speaking, cultural revolutions.

The first revolution, necessarily, is the originating one—the one which creates technology, which separates the life of technological practices from organic, animal life, and which, in this sense, speciates *homo sapiens sapiens*. Thus, the first revolution introduces the very creation and use of tools, and of the methods of their making and use. Both the dating and the characterization of this revolution are hotly debated issues in contemporary anthropology, and I will not discuss the issue here beyond remarking that it depends on the dating and interpretation of archaeological artifacts *as* tools. We read backwards here. Because we have the category "tool" defined more or less clearly in terms of what we recognize as later and familiar exemplars or criterial cases, we attempt to recapture or reconstruct the germinal forms of such classical tools. The tool, however, is not simply an object, but a function which defines an object as the means by which this function is carried out. Some object is a weapon—a spear, an axe, an arrow, a club—because it can be used to perform this weaponly function. More teleologically still—if we are talking of tools in this direct and simplistic sense—some object is a tool if it has been made for the purpose of such a use. We have, therefore, the class of use-directed teleological artifacts, embodying conscious and deliberate human purposes, and the technical knowledge required for carrying out these purposes by means of auxiliary additions to the human body. The precondition (or co-condition) for such artifact-production is the means of social cooperation and direction by which these practices or these activities can be shared, preserved, and transmitted, among humans, and between generations—i.e., some language or form of communication. Language, John Dewey said, is "the tool of tools," not in the sense that it is simply the most complex or highest means of such life activity, but because without it, the social activity of the making and use of tools itself could not be carried out.

Such primary tools, roughly speaking, are hand tools—or perhaps *body* tools, if we include such things as harnesses for pulling, etc.—instruments used by the person as extensions or adjuncts of such bodily actions as pushing, pounding, pulling, lifting, striking, throwing, digging, spinning something around, or piercing something. Some other distinctions may be drawn here, among such originating tools: (a) that there is a class of tools used to make other tools, e.g., the pounding stone, in flake and core production; and (b) a further class of tool models or prototypes from which other tools are made as copies. This also introduces the class of those tools, divorced from their primary use in production, which function as representational artifacts, which become the first symbolic tools, or images, and indeed, the first embodied universals, as models for reproduction.

This technology of tool use and tool production has as its correlate the technologies of social organization, of norms of practice, cognitive and linguistic methods of preserving and transmitting the techniques and rules, therefore the representational and expressive technologies that involve what we characterize as ritual and magic.

In this hasty sketch, these aspects can only be alluded to. The more general point is this. Every so-called technical change, every invention or introduction of a tool, entails a host of social, organizational, and cognitive practices: the production or reproduction of tools, the preservation and transmission of these acquired skills, or techniques, the forms of collective or collaborative activity required in such making, and in the coordinated use of hunting tools, or agricultural or domestic implements, the stresses and requirements put on language by such changes, and by the modes of cultural education that they require. All of these constitute what, broadly speaking, and by analogy to Marx's analysis of production, we may characterize as the social relations of technology, in distinction from the technological forces, the material and conceptual means or instruments which function within these relations. However, there is no suggestion here that the material forces of technology somehow "determine" the technological relations uniquely, whatever "determine" may mean here. It may be, in fact, just the reverse: that a certain existing set of social relations is receptive to, and encourages the development of a certain kind of productive artifact, or tool, or calls forth or stimulates sheer invention, or the imaginative play of innovation. (What Ellul calls the "mystery of invention.")

The first revolution, then, is the revolution of the hand tool (or body tool), by means of which the ordinary organic bodily skills and actions, the uses of human force and wit by direct application of mind and muscle, is extended, enhanced, or replaced by an artificial—i.e., artifactual—means.

Two important features of this first revolution are worth noting here. *First*, the difficulty *we* have in recognizing the extraordinary complexity and scope of the achievement of the simplest primordial tools. This complexity may be gauged in part by the extraordinary (in our terms) length of time—the tens of thousands of years—that it took for a species no less biologically equipped than we are, to invent, elaborate, and transmit the culture of tool production, indeed, to preserve it for such long periods with little change. The transition from biology to culture, from animal to human life, is therefore the longest revolution, with its own inner stages—a complex phenomenon in which, for a time, initially at least, cultural practice enters as an element in evolutionary biological adaptation, as the forebrain or cerebral cortex, in its development, internalizes into genetic structure, the growing dominance of tongue and hand, of speech and tool making and tool use in the shaping of hominid life forms; and as bipedal locomotion and the opposable thumb free the hands for these uses and others.

In this development, the making of images comes very late—only about 35,000 years ago—and with it, the possibilities of the visual imagination, a cognitive revolution so radical that its measure remains to be taken and understood.[1]

Second, the *later* development of tools is therefore *culturally* evolved, built on previous achievements—therefore substituting a Lamarckian for a Darwinian mode of evolution, involving the transmission of acquired characteristics from one generation to the next, and indeed, even from children to parents, in a reverse mode.[2] Cultural evolution is, by contrast to biological evolution, explosive. Each generation is able to transmit its acquired characteristics—its know-how, invention, skills, experience—to the next, without the much slower multi-generational mechanisms of natural selection. The cumulative cultural achievement, what Dewey call "funded experience," therefore serves as the basis of new innovation, of the proliferation of alternative possibilities for trial and error selection. Thus, an authentically *historical* process is introduced for the first time, beyond the limits of so called natural history—a cultural rather than a biological history; and now one created by human choices and actions, by cognitive practices.

It is appropriate to call this first revolution the *Promethean* revolution, since it is the Prometheus myth that first fully expresses the human self-recognition of the nature of this revolution. Prometheus stole fire from the gods and gave it to humans; but also the arts of production, of agriculture, of building, etc.—in short, the tool culture.

The second revolution is an ancient one, by our historical reckoning, though it is realized fully only recently, in what we call "the industrial revolution." But the technological revolution which lies behind or provides the technical means for this industrial revolution is much older. It is the complex transition from hand tool to machine, i.e., to the "tool" that elaborates a process of production, or a sequence of actions of hand tool use, in such a way that the repetition or duplication of this action does not require the repetition or duplication of a human skilled action. The machine is essentially a device for embodying or objectifying a human skilled action in a material design which (a) performs an effective imitation of or substitution for that action by itself and (b) can repeat the same performance many times. The machine therefore puts the application of human skilled force or work at two or more removes from the point of production, and substitutes for it a designed mechanism which exhibits, in objective material structure, the cognitive understanding of the production process as an abstract process—i.e., in terms of its formal structure. Thus, every machine is a formal realization of a function. In this sense, machines are characteristically platonic objects: formal structures or designs, represented in blueprints, and instantiated in material objects which are "copies" of the design. The actual *performance* of the machine is therefore the actualization of the function which the formal structure is designed to realize. Now, one may say that the same is true of hand tools; prototypes, models, rules of construction all serve as formal models—whether embodied externally, or as sets of instructions, or visual images "in the mind" so to speak—i.e., as internal representations. The "platonism" here is incipient in the idea of *making the same*,[3] i.e., reproducing or duplicating. With the machine, however, a much more radical abstraction takes place, perhaps even by stages: *first*, when a manual or other bodily skilled practice is transferred imaginatively to a functional analogue (e.g., hurling a projectile by hand to hurling it by means of a sling or a bow; or grasping something by hand, to grasping it by pliers or a wrench); and *second*, when a complex sequence of actions is analyzed and reconstructed in a connected structure made up of continuous parts which act on each other. What has been achieved then is the typical causal structure of a *mechanism*.

In this more radical abstract reconstruction of a work process, the connection to concrete handwork, the imaging of a direct analog may be transcended, as the work comes to be imagined in an autonomously mechanical way, as invention replaces imitation, and as practical, constructive metaphor—seeing one thing as another—frees the imagination for creating new ways of working, new techniques of production, new forms of appropriating natural forces. From this mechanization of work, and the analogical mechanization of analysis (of a complex whole into its parts) and of synthesis (the reconstruction of the whole from its parts), a change in the representation and understanding of nature, as well as of social existence and the self-understanding of consciousness as *rational* consciousness, takes place.

The world is reconstructed as a machine, as is the human body, and the social structure as well. The analogue to geometrical and arithmetic construction is realized not only in the machine, but also in the world machine. The geometrization and quantification of nature, as well as of human nature, takes place, and in rapid succession we have the mathematization of motion, now understood as mechanical change of place, in a coordinate space divided into its equal unit parts, whether in the cartographic grid, or in the pictorial representation of the visual world by the projective geometrical constructions of linear perspective, or in the later Cartesian coordinates.

The second or machine revolution is profound, then, not alone in the transformation and productivization of handwork as bodily skills, but in the transformation of cognitive mode, i.e., of the very basic norms and structures of knowledge, both in the understanding of nature and in human self-understanding. Human beings come to understand themselves in the model; and in the sense that, as cultural beings, we are in significant part what we take ourselves to be, humans are in this sense self-constituting and self-transferring. This second revolution also brings with it an historical change in the character of human and social reality, as well as in the modes of cognition.

Thus, the first feature of this revolution is the transition from handwork to machines, with all of its concomitant transformations deriving from the idea and practice of mechanism. The second feature of the machine revolution is that, in addition to radically increasing the scope, speed, accuracy, and output of former handwork, it also makes possible the application of other sources of power or work than the exertion of human force. Draft animals represent the simplest forms of such substitution, in agriculture and transportation, and in rotary machines (e.g., water pumps, grinding mills, hoisting pulleys). But then, in rapid sequence, wind, moving water, heat, and explosive chemical reactions are harnessed to machines as sources of power, for sails, windmills, steam engines, combustion engines. And at the acme of this revolution, the generation of electrical energy and the transmission of energy at a distance marks the universalization of the mechanical revolution, now joined to thermodynamic and electrochemical sources of power. Atomic power, at least in terms of fission if not yet in terms of fusion, extends the range of energy sources for this mechanical revolution, but does not as such change its character.

What changes the character of the second revolution and introduces the third revolution is the use of calculating or computing machines *in conjunction with* other machines, as a means of control of the machine processes automatically. In short, the revolution comes with automated production by means of information and control devices—at first mechanical, then electronic.

This third revolution is, in one sense, only an elaboration of the second. Calculating machines are still machines. When electronic connections replace mechanical ones; when linear sequences are reconstructed in circuit design; when combinations of combinations of simple circuit connections are integrated in a complex switching device; when the time of sequential "actions" or calculational events proceeds from the molar measures of mechanical motions to the nanoseconds of electronic switching devices to the trillion-operations-per-second teraflop computers of the near future—though the mechanism remains a mechanism, new qualities emerge even within the mechanical or electrochemical domain. In any case, the machine has now become electronically controlled. Human design now embodies the administrative and operator functions in a program of instructions, i.e., in an objectified rationality capable of operating by itself once it is switched on, so to speak. Fully automated production, in this sense, is simply substitution of a program of automatic commands for the program of human commands of an earlier production mode.

There is one other major aspect of the third revolution which has sometimes been taken as a revolution in itself; that is the close relation between theoretical science and applied science in this period. This works in two ways: one, obviously, is the rapid translation of theoretical principles into practical applications, often by the organization of deliberate mission-oriented crash programs. A classic case is the Manhattan project—the design, building, and testing of an atomic bomb. Another, more recent example of the principle at work is the demand made, in recent discussions of so-called industrialization policy, for greater emphasis on product realization and process engineering—classically "downstream" engineering functions of low priority in engineering education—and more government funding for research and technology for economic development, less for basic research.[4] The second way is the enormous dependence of contemporary theoretical science on high technology, not only in the exotic precincts of the design and construction of supercolliders for high-energy physics research, but in the everyday contexts of experimental measurement and data analysis—e.g., in the life sciences and medicine, and in chemistry in all its branches. One look at the annual instrumentation guide for biomedical research, published in *Science*, staggers the technological imagination, showing how fundamentally dependent theoretical science is on measurement and calculative technologies. The most abstruse theoretical physics, as well as the most applied design engineering, depend on supercomputers; and both are screaming for more and faster devices for their basic work. The supercomputer race between Japan and the United States now becomes a central issue of international power—as paradigmatic a fourth revolution issue as one could wish.

This overly simplified sketch of the three revolutions is extremely onesided. It describes only the narrowly technological character of these revolutions in terms of the dominant means of production, with no elaboration of the characteristic methods in each case, nor of the characteristic social and cognitive foundations of each stage. On the other hand, if we are here to consider the fourth revolution usefully, it cannot be described in this *narrow* technological sense. For, in these terms, nothing beyond the third revolution occurs here. Rather, it is both the quantitative and qualitative features of the third revolution, as it develops in its latest stages, that generate the fourth revolution as a revolution—not in technology, but in the relative role or place of technology in the broader contexts of culture and politics.

What is it about the third revolution that leads to this consequence? There are several obvious and some non-obvious factors. Among the obvious ones is the extraordinary increase in productivity, measured in two ways: first, by the ratio of unit cost to capital investment plus labor costs (or fixed plus variable capital costs, where capital investment includes costs of plant, machinery and other fixed costs); and, second, by the ratio of unit cost to labor cost alone (or variable capital costs alone). The first measure shows the economic power that results from lowered unit costs; where there is a market in a given commodity, the lowest unit cost of production of that commodity (other things being equal—e.g., quality control, no differential or protective tariffs, no monopolistic practices, etc.) will dominate the market. The consequence of the second measure—i.e., the ratio of unit cost to labor costs alone—is that, other things being equal (e.g., capital costs are in the same high technology), the economy with the lower labor cost will dominate the market. These are crude—i.e., "pure"—economic phenomena in an ideal "free competitive market." But, given these variables, the stakes for market domination become clearly tied to the most advanced production technologies and the most controlled and stable labor markets at the lowest labor rates. But variables can be manipulated here, for example: rates of financing of capital investment, or credit rates; tax rates on profits; imposition of protective tariffs on competitive imports; labor legislation, labor-regulatory and strike-regulatory laws; minimum wage and maximum hour laws; level of government support or funding of basic scientific research, and of major high-technology research and production as this affects technological superiority; availability of skilled engineering and administrative personnel and of a skilled and technically competent labor force; and, of course, the stability or growth of demand, as itself a manipulable variable within the limits of available purchasing power.

This is only one side of the situation as it relates to high technology in the production of economic goods. It is simple enough to see the analogous situation with respect to military superiority. Here, in connection with the more general context of a technologically superior economy, there is the matter of direct subvention through military budgets of the most advanced technology out of public funds. These in turn become matters for policy decision both within the military (as to priorities and requests) and in national government, which taxes for those funds and allocates them. The strongly interlocking set of interests here—in government, the military, and industry—makes the issue of technological power and development as clearly a national political question as one can imagine.

Another obvious effect of the third revolution which generates the fourth is the sheer environmental effect of the third revolution. Small symptoms go back centuries. The blackening of the English cities and countryside, and the smog-ridden "dark, satanic mills" of Manchester (as of Pittsburgh), in the older days—when daylight was blocked by factory soot, and lungs were coated with coal dust. Earlier, the denuded slopes of the Adriatic, where over centuries timbering for shipbuilding turned forested mountains into scrub and rock surfaces. There is, here, the typical litany of local environmental pollution resulting from industrialization and unregulated waste disposal of the second revolution. But now we have global effects in prospect, some of them prospectively irreversible and life-threatening for the planet as a whole: the Chernobyl disaster; the greenhouse effect; the ozone hole; acid rain; nuclear waste disposal—and the evidently lessening but still persistent threat of nuclear war and its fall-out consequences including nuclear winter.

The reason for including even so bare an account—along with such an obvious list of the economic, military, and environmental consequences of the second and third revolutions—is to characterize some of the objective conditions of the fourth revolution, namely, those conditions which politicize technology as a central question of national policy, the national economy, international competition, rivalry, or war, and governmental or global regulation of massive hazards for species life. All this is new. This does not mean that aspects of such problems did not already show themselves much earlier. But in the fourth revolution, as in the previous three, there is no sharp demarcating or exclusive characterization of the features of the revolution, as if they simply replaced all the others. Hand tools of the earlier sort are still in use, ubiquitously, as are older sorts of machinery, side by side with robotics and automated production. The hand lathe in the local machine shop is cousin to the automated multi-tool turret lathe in a computerized production system, and both are in operation. There were capitalists and proletarians in ancient Rome, but it was not thereby a capitalist economy. What concerns us is the dominant mode of technology, that which does not simply displace all the others, but which, comparatively speaking, trivializes them, makes them marginal or vestigial, in practical social-economic and technical terms.

The measure of the fourth revolution's dominance is the degree to which high technology has become central to the economy, to government policy, to the military, a life and death issue of international hegemony, of competitive edge, of military superiority or inferiority. This is, of course, tautological: what I mean by the fourth revolution is just this dominance of technology in national political and economic life.[5]

POWER

The issue which emerges, then, is the control and direction of technology in relation to *power*—economic and military but ultimately political power. The notion of power, in these terms, deserves some analysis. First, we may distinguish the power *of* technology—technological power—from the power which control over such technology affords. A simple example, from the first revolution, will make this clear. The technological power of the bow, as a weapon, is the speed, force, and accuracy with which it can deliver an arrow—i.e., a piercing weapon—across a great distance (compared, of course, to alternative projectile methods such as throwing or hurling). Its technological power is therefore a comparative measure of its deadly force as a weapon, given available alternatives. The power which control of such technology affords, however, is the military power of subduing and killing enemies, the social power of controlling the hunt, or the economic and social power of provisioning the tribe more adequately, or in conditions of scarcity or of competition for game, advantage over others without such means.

In contemporary terms, one may also distinguish between the technological power that atomic fission provides, in terms of energy, and the power which control over such an energy source would give. This was the burden of Albert Einstein's letter to President Roosevelt, during the Second World War, concerning the need to build an atomic bomb before the Germans could succeed in building one. The scientific theory of nuclear fission predicts the energy released by fission; the technology makes a device which can release this energy under controlled conditions. Politics and military policy determine the use of such a device to devastate an enemy and to win a war.

It is this latter, political power utilizing the technology, that is of concern here in the analysis of power—not technological power (which deserves its own analysis, but not here), then, but rather power in the use and control of technology. Here, some elementary considerations to start with.

(1) To control a technology means to be *able* to use it at will, i.e., to know *how to use it*, and *how to cease using it*. Thus, in the case of typical third-revolution high technology, control requires scientific and technical knowledge and skill adequate to the sophistication of the technology. Conversely, in the absence of such knowledge or skill, the force of a technology cannot be utilized or controlled.

(2) To have power means to be able to exercise it *freely*. Thus, if I can exercise a certain power only upon direction or decision by someone else, the power is not mine; I am simply an instrument or administrator of that power, and not its agent. Power, thus, entails *agency*, or freedom to act.

(3) Power, as effective control over a technology presupposes that the technology is effective; otherwise control over it is empty. Control over and use of an ineffectual technology is not power.

To summarize, agency without skill is blind; and without force, empty. But these rather obvious assertions make for an interesting consideration of power in relation to technology. Actual decision-making in the deployment of high technology may very often be a technician's or engineer's prerogative, but only within policy set by others—e.g., a corporate board, or a government agency. In this sense, though the experts may ultimately be needed to *use* the technology, having the requisite skills, they are not free to use it, and thus have no power over it, or exercise no agency.

Another consequence: I need not control the technology by myself to be an agent; if I share equally in the policy decisions, then I have shared a common agency with others, and am free in this sense, as a social agent. If, however, I do not have the skills to make decisions on the use of the technology, I am also deficient in agency. Therefore, optimally, control over technology, hence power, really depends on possessing the requisite skill and at least shared agency. Of course, if I have the requisite know-how, and am free to make policy decisions with respect to an effective technology, I may be said to have power. The two conditions, then, are *knowledge* (technical-scientific knowledge and skill) and *agency* (or authority to act).

From all this it follows that the power of technology may be vested in an individual or in a group, but only where agency and skill are combined. But now as in considerations of power more generally, one may distinguish between *power*, as effective use and control over technology, and *authority*—as *rightful* effective use. Not all exercise of power is rightful; thus, the use of power may lack authority. What, then, would make the exercise of power in the contexts of high technology normatively justified or authoritative?

Here I cannot get into the analysis of such normative justification, but I can outline the problem to be addressed.

At present, effective policy decisions on technology are made by various corporate or government or military bodies generally organized along strongly hierarchical, if not authoritarian, lines. Sharing in the decision making consultatively are the technical experts who advise policymakers, though they do not make policy themselves. In government contexts, in parliamentary democracies, appropriate committees (e.g., congressional subcommittees, in the United States) make recommendations for congressional action and ultimate approval by the Executive. Hence, there is a modicum of democratic representation in such decisions, though it is indirect. One may say that those most likely to bear the consequences of a policy decision should have something to say about it in a democracy; and that democracy is the favored form of government. This is not an argument for democracy but a stipulation or a premise.[6] But two problems arise here. Politics being what it is, democratic procedure in reaching decisions does not guarantee the best or the right decision; and it is reasonable *not* to accept the view that a decision is the right one simply *because* it has been arrived at democratically. (A majority voted to condemn Socrates to death, in an extraordinarily democratic judicial system, but they were wrong. And Hitler became chancellor by legal parliamentary means, after the party won the 1932 election by a plurality.) The right decision is presumably the most informed, best reasoned, and competent one in the circumstances. At a minimum, this takes knowledge or an understanding of technology. Again, the technicians are the consultants, in matters where high sophistication in scientific and technical understanding is involved, as it is in many contemporary contexts. But to judge their counsel itself requires some nominal technical understanding of high technology and its consequences. Ideally, this would require a technically and scientifically informed electorate.

Thus, the two conditions for democratic participation in the control and use of fourth-revolution high technology are: (1) adequately democratic modes of sharing or participating or being represented in decisions; and (2) adequately informed understanding of scientific-technical questions. These are the basic political and educational conditions for the democratization of the fourth revolution, a democratization devoutly to be wished, but without guarantees of right decision even if achieved.

TECHNOLOGY AND TRUTH

If technology's relation to power is obvious, its relation to truth is not. In part, this is because technology's characterization as instrumental reason locates its normative structures within the contexts of successful or unsuccessful action, or effectiveness, or adequacy, or economy of means. That some process or instrument works well to achieve a given end, or is the most efficient or economical means to an end, has to do with the virtues of successful practice, not of truth. In only one context do the two coincide: in the pragmatic criterion of truth, where the veridical is, by definition, or by philosophic argument, identified with the practical. "Give me a truth I can ride," says William James. "The ultimate test of truth is *praxis*," says Karl Marx. But, clearly, a method or a tool, a way of successful action, is not "true" or "false" in the sense that statements are "true" or "false"; and whatever difficulties attend the correspondence or coherence theories of truth, a technique is not true or false on the grounds of its correspondence to anything, or on the grounds of its coherence with other techniques.

There is another sense, related to the pragmatic *test* of truth, in which technology is instrumentally related to truth—and that is as a means for testing hypotheses or theories. Technology as experimental practice, or the technologies *of* experimental practice—technologies of measurement, or technologies of experimental design and experimental apparatus—all of these play a role as means in the process of testing the truth claims of science. Given the profound differences among philosophers about the very nature of truth claims in science—whether in the older or in the current debates on instrumentalist vs. realist interpretations of scientific theories, or between realism, irrealism, conventionalism, and constructivism—we can hardly look here for the resolution of our issue concerning technology and truth. Where, then, shall we look?

Traditionally (and correctly, I believe) truth has been characterized as a function of propositions of a certain sort: those that refer. Truth was said to lie in the correspondence between what a proposition asserted and the state of affairs designated or denoted by that assertion. Truth was not "in" things, but in language or in some relation between language and what language is about. The notorious difficulties for which the semantic theory of truth is famous warns us that glib, simple, even intuitive approaches will not get us very far philosophically. (Recent "minimalist" semantic theories of truth tend to radically deconstruct the older game and opt for very intuitive solutions.) None of this, however, is what I am concerned with here. Rather, I am proposing an alternative, a facing away from the traditional problem, and an approach from the direction of what I have called Impure Reason[7]—where critique is essentially a reconstruction of technological modes of thought.

The gist of this is, briefly: Technological modes of thought are concerned with reflection upon and critique of technological praxis—of the actions by which things get made, or organized, or changed, or destroyed, or put to use, or improved, redesigned, taken apart and put together again. We learn two things in the course of such *praxis* (or *poiesis*, strictly speaking). First, we learn about the things made, unmade, remade, in the same basic ways that children learn by manipulating, constructing, deconstructing things. We learn what can be done with what, to what, up to what point, and with what effect. We achieve what Giambattista Vico called "maker's knowledge": "We know what we make," he said. But there is something else we learn in this activity. We find out what we can make, what we can do, what our capacities are. We enlarge our capacities in the practices we elaborate and revise; we *become* competent and know ourselves to be so; or we discover our incompetencies and our limits. In short, we come to know ourselves, not as we *are*, in some fixed and static image of the inner self, or essential nature, but as we become and change. Our knowing ourselves is itself part of what constitutes our being ourselves, so we construct ourselves, and review the construction.

Our activity is not solitary. So we know our social existence, our being in and through others. None of this is new. It is Socrates, Vico, Hegel, Feuerbach, Marx, Mead, Freud, in various aspects. But the beings that come to know themselves, in and through their own activity, are just these technological beings who, in producing the means of their existence, also produce a world out of the given stuff (or the *taken* stuff) which is their world in and through which they know themselves. This access to world and self through constructive activity—not of the "mind," in some disembodied and ghostly epistemology—but of the minded hand and foot and heart—this access is access to *made* truth, to what is made true, and truly made, in a constructive correspondence whose objective constraints are the recalcitrances, resistances, refusals of the unmade world, consenting to be made only on its own terms, to be discovered by the makers, or imagined or dreamed by them.

"True" is *treu, getreu*, dependable, reliable, like a true friend, before the concept gets caught up in mirror metaphors and correspondence theories of truth. "True" as a term of art (or as a term of *act*) has a history and a genealogy that goes back to technology, to the simple "truing up" of the carpenter, to the "rightness" of a right angle, to the rectitude of the good person. This is no Heideggerian game of etymological poetics, but of historical semantics. Technological "truth" is the truth found to be true, from which, then, assertions may be generated to celebrate and to communicate the discovery. *Then* we can have "true" sentences too—at least for as long as they remain *treu*: loyal, dependable, reliable.

The fourth revolution, by contrast to the first three, introduces a terrifying option; it makes technological or maker's truth hostage to political power, in a decision-procedure that tests policy against the lives of millions, against the planet's future. Now, a policy decision is not itself a truth claim—it typically proposes that something be done to achieve a certain end. But it is based on truth claims, on presumptive knowledge of the facts of the matter. However loose the fit between intentions and outcomes in policy matters, good faith requires some reading of the relevant facts, in their best determination, upon which the policy decision is crucially based. The willful distortion or suppression of facts, or even of reasonable conjectures and arguments about the facts, in the interests of some favored policy goal, or of some exercise of power, is the most dangerous corruption that the politicization of technology makes possible in the context of the fourth revolution.

In testifying before a U.S. Senate subcommittee studying the greenhouse effect, James Hansen, director of the Goddard Institute for Space Studies in New York, had his report on the greenhouse effect tampered with by the Office of Management and Budget. The purpose was to minimize the effect of his view that the computer models which project a gradually increasing curve may depart from the gradual curve and suddenly become far more pronounced. Because such a view was seen to be at odds with the Administration's line, OMB edited his testimony. I quote from the report in *Science*: "Denouncing the White House Intervention as censorship, Senator Albert Gore (D, Tenn.) said 'The Bush Administration is acting as if it is scared of the truth.'" What is at issue here is not some hard and fast truth, but a scientific conjecture, based on a technology that is an integral part of modern theoretical science: computer modeling of the weather. Hansen's view may well be wrong, but it may well be right. The point is, policy on the greenhouse effect needs to be well-informed, as a typical fourth-revolution issue of a globally politicized technology. Whatever the merits of Hansen's view, it cannot be censored or deformed on grounds of its political unpalatability without undercutting the reliability of any policy decision.

The dangers of uninformed, underinformed, misinformed policy decisions on issues of this scale and scope are so great that truth, or the best available versions of it, are requisite for decisions on the rational control of technology.

The power over technology in the hands of those who are afraid of the truth is the most dangerous power on earth. The imperatives of democracy, and of technological-scientific literacy, seem to me to be the most promising antidotes to such power.

Baruch College, City University of New York

NOTES

1. See Randall White, "Visual Thinking in the Ice Age," *Scientific American* (July, 1989), pp. 92-99; also, A. Leroi-Gourhan, *Le Geste et la parole*, 2 vols. (Paris: Michel, 1964).
2. On the view that cultural evolution is Lamarckian in contrast to biological evolution, which is neo-Darwinian, see my discussion in "Sight, Symbol and Society: Toward a History of Visual Perception," *Philosophic Exchange* 3:2 (1981): 28; and also in "Perception, Representation and the Forms of Action: Towards an Historical Epistemology," in my *Models: Representation and the Scientific Understanding* (Dordrecht: Reidel, 1979), pp. 204-205.
3. See Ernst Fischer's discussion of the role of "making alike" in the origins of art and of cognition, in his *The Necessity of Art*, trans. A. Bostock (Baltimore, Md.: Penguin Books, 1963), pp. 29-33.
4. Both of those recommendations are from the MIT Commission on Industrial Productivity chaired by Michael Dertouzos as reported in "Toward a New Industrial America," *Scientific American* (June, 1989).
5. A brief review of the fourth revolution's vector can be gleaned from a sampling of reports in *Science* within a period of a few months: "Inaction on Technology Stirs Congress" (14 April 1989); "Global Supercomputer Race for High Stakes" (14 April 1989); "Skeleton Alleged in Stealth Bomber's Closet" (12 May 1989); "HDTV" (19 May 1989). There is nothing surprising in this, but its very familiarity masks the increasing politicization of technology—side by side, one may add, with the technologization of politics.
6. For a recent normative-critical discussion of the question of democracy and technological control, see Carol Gould, *Rethinking Democracy* (New York: Cambridge University Press, 1988), chapter 10.
7. See M. W. Wartofsky, "The Critique of Impure Reason II: Sin, Science and Society," *Science, Technology, and Human Values* 6, no. 33 (1980): 5-24. "Critique of Impure Reason" was used by me earlier in the title of a different paper; the phrase, obvious enough as a play on Kant's title, has apparently been used since without attribution, but innocently. Ludwig Feuerbach had, already in 1841, considered the title, "The Critique of Pure Unreason," as one of the options for his work, finally entitled *The Essence of Christianity*.

JACQUES ELLUL

TECHNOLOGY AND DEMOCRACY

Our society now faces an essential challenge. I will not attempt to exhaust all sides of the multifaceted debate about technology and democracy, but will instead summarize my present thinking on the topic.

In the course of my work, I have written little about democracy. For my purposes here, I shall not deal with a host of traditional questions, such as whether television, information science, the mass media, and the global village are favorable to democracy, whether they facilitate or undermine democratic practices. Neither will I discuss the problem much bandied about in recent times concerning the democratization of technique, the problem of whether people can control technique and harness scientific research. I will also leave aside another much-debated question: Is technocracy in the process of establishing itself, co-opting traditional politics?

In my view, our Western political institutions are no longer in any sense democratic. We see the concept of democracy called into question by the manipulation of the media, the falsification of political discourse, and the establishment of a political class that, in all countries where it is found, simply negates democracy. While such problems do not stem from technique alone, there are strong indirect links; techniques of organization, management, and the control of public opinion help maintain the political class in power. To understand democracy's predicament, however, one must look deeper. I shall view the matter from the standpoint of three periods or ages, a perspective I have outlined in *What I Believe*.[1]

THEORY OF THREE AGES

By "milieu" or age I mean the predominant human environment that furnishes mankind with all that is needed in order to live but that, at the same time, is the cause, source, and origin of the greatest dangers, what Toynbee calls "challenges." I believe that in the course of all human evolution there have been three successive ages of this kind.

The first period can be designated "prehistory," in which the natural milieu is dominant. Without seeing nature as either a value or a concept, one can notice that from about 500,000 to 5,000 B.C. man is closely bound to nature, a condition which, from pole to equator, requires adaptation for survival. In this era humans find nourishment and protection from what nature offers. Mankind lives first from gathering and then from hunting, consuming, and eating what the natural environment provides. Everything is received from nature, including the first tools. Human beings live in widely dispersed bands, requiring vast spaces from which to exact their sustenance. But nature is also the source and the sole cause of the dangers at this stage—dangerous animals, famines, floods, and so on.

Once a certain population density is reached, there are more frequent contacts among groups; people come together in a variety of unions and exchanges and soon the first cities appear. This is an important turning point, for within the city, man finds nothing "natural" to sustain life. Although people still depend on an external natural world, they proceed to set up a new order characterized by expanding relations among groups, increasing exploitation of natural resources, and the rise of a rudimentary division of labor. In other words, a second order is formed for humanity, which, in the course of the "historical" age becomes society in its true sense. The human being now lives within an organized society. The sharing of labor and responsibility, coordination of collective power, cooperation for defense against natural dangers, and creation of rules for coexistence are now the conditions that make life possible.

Of course, the natural realm still exists; from it alone the means of existence are always drawn. But a natural order tends to be subordinated, subjected, and rearranged. The more progress is made the more this order ceases to be "natural," integrated as it is into organization and culture. As a corollary, we can see that starting from the point where humankind only subsists on what nature offers, we arrive at the stage where a subordinated nature produces for human needs. Natural dangers (storms, wild animals, etc.) continue, indeed, but in a more limited way, as society finds more numerous means of protection against these challenges. At the same time other dangers appear, ones that again put people at risk, but these are "social" dangers —war, struggles for control, slavery, etc. Within social groups at this stage of development power always arises, power which can be either religious or political in character. Thus, politics (*la politique*[2]) can appear only when man is plunged into the social order. As politics grows in importance within the social order, there are attempts to normalize power through "democracy." These steps confer power on the "people"; that is to say, on those with the most at stake. This is surely wise, but experience shows how difficult it is to implement democracy, a point I shall address later.

Eventually, the human order changes yet a third time. Beginning in Europe around the eighteenth century, accelerating in the nineteenth century, and reaching maturity about 1950—the order people confront becomes Technical.[3] Populations are grouped in giant cities, a step made possible only through innumerable techniques. Henceforth, all the possibilities for human life are drawn from technique alone. Social organization itself now depends precisely on a growing number of techniques. At the same time, the dangers that threaten humanity now come from these very techniques, dangers that are material, psychological, or relational. In this period the old phenomena of nature no longer have the meaning they once conveyed. Of course, the natural order still exists, but it is totally dominated, manipulated, and, in the worst cases, faced with extinction. Nature still creates havoc, but today we must question seriously whether some of those catastrophes (drought, warming effects, cyclones, and changes in the monsoon seasons) are not caused by an increasingly aggressive human interference with the natural order of things.

The social order and society continue to exist in our time as well, but they tend to collapse as the dependence on technique spreads. It is no longer possible to find a common measure between the old phenomena and present-day phenomena. One cannot, for example, compare an eighteenth-century war with holocausts produced through twentieth-century technical means. Furthermore, all that makes social groups possible—myths, beliefs, laws, morality—is reduced to nothing by technical growth. For technique has become the paradigm for all action, furnishing to any organization or process both its ontology and its underlying logic. Thus, as we have seen, the political expression of order in "society," has practically ceased as the primary organizing force. For the technical order requires the subordination of social and political power whose practicality, in any case, appears less and less evident. If we term the period from 5,000 B.C. to 1900 A.D. "historic," we can say now that we are living in the "posthistoric" era.

If this is true, does it make sense to talk about democracy and technology in the same breath? Does democracy as a means of political organization still have a place and a value? I would respond "yes," without hesitation. I would add, furthermore, that it is the decisive question to ask ourselves. The stakes are high: it is a matter of knowing whether the force of things will henceforth dominate human evolution and every individual, or whether human will (that is, the will of the individual as a human being) "still" and "again" has the power to influence the course of events. Will the future be limited to the linking of techniques into a system within a purely mechanical type of evolution? Or, will some form of history yet prevail?

If technique is the true force of things, then we will surely witness an end to history. In this case this conflict will become the final one in which the technical solution becomes irreversible. As Cornelius Castoriadis has pointed out, technique in its full development is a closing, whereas a true democracy is an endless opening. Thus, in the choice before us, the democratic option is not sentimental, ideological, or moral; it is not just one among many other alternatives. In reality, there can be no other alternative for man.

If the present course of events continues, all forms of monocracy (monarchy or dictatorship) will inevitably be dominated by the technical function. All forms of aristocracy (bureaucracy, *nomenklatura*, etc.) will turn themselves quickly into technocracy. Power will move steadily toward technocrats, the true aristocrats of our society. From there on, the alternatives are clear: either technique or democracy. In the absence of democracy will any person still have the possibility of being human?

In the natural and societal orders that preceded the technological age, humanity was always able to derive profit and advantage from a specific order, while also avoiding its disadvantages. The same possibilities present themselves to us in the third age. I wish to be very clear that in this sequence of events, man did not destroy nature during the first period; neither did he destroy society in the second period. Nor is it a concern in our period to destroy technique. But the major difference between our periods and those preceding ours is the fact that, in those periods, man did not have the power to destroy society or nature, a power to specifically dominate or exploit those orders. With technique the situation is completely different: power has become technical and technique exercises a pervasive form of power. Hence, to achieve democracy would require abandoning the technical system in its totality.

But is this possible? Given the enormous risks, both actual and potential (of which we now know a small part and which will reveal themselves as we go along), a new mode of democratic organization must seek to limit risks and dangers. At the same time it must fine tune the utilization of techniques. Hence, the question for democracy arises: can the people actually exercise power of this kind?

One must immediately take account of the fact that two different logics demand obedience: namely, the logic of efficiency and the logic of democracy, the logic of the quantitative as against the logic of the qualitative.[4] Democracy is not and cannot be efficient as long as it is truly democratic. The logic of efficiency concerns solely the efficiency of process and never the efficiency of results. Technique is not fundamentally efficient if one measures the totality of its results—waste, pollution, external costs, excess production, planned obsolescence, which requires throwing away perfectly useful objects that are replaced by higher performance models, quasi-food produced by chemical additives or techniques of the feed lot, the ephemeral character of the majority of the objects that technoscience offers us as a result of research and development.[5]

Duverger's recent book on the dictatorship of impotence has quite unintentionally thrown light on this contradiction.[6] Duverger stoutly defends majority rule against proportional rule. Proportional rule, he argues, breaks down political organizations into a multitude of small parties, where one governs only through alliances and coalitions that are always fragile. However, in his view, government must account for all tendencies in society and will collapse on the first occasion when it is not at all free to carry out coherent policies that are strong and uncompromising. Under majority rule one ends up with two parties and the one which obtains the majority can govern in peace and carry out its projects.

This argument is irrefutable, but it is based on the primacy of efficiency, the technical criterion. Duverger does not realize that what he proposes is precisely the denial of all democracy. Indeed, such a system imprisons public opinion by forcing black and white choices. New opinions are often excluded by being reduced to two viewpoints. In such a political system the only possibility is endless repetition.

Politics of this kind also excludes diverse minorities. It is, however, a fundamental characteristic of a democracy that minorities should be considered and able to express themselves, that they should even be considered privileged. Any system that reduces choice to two candidates also limits the possibility of debate. What happens when an election gives one the choice between a dishonest candidate and a stupid one? In the last analysis, everything that makes a democracy a human system is eliminated if one applies the criterion of efficiency.

The question of technique versus democracy presents us with a stark conflict of power. Thus, we must rethink democracy and form a new concept of it, drawing upon the ancient model. We must recognize, first of all, that representative democracy has failed, both politically and juridically as well as socially. Representative democracy might have instituted a form of power that the people could exercise effectively over their representatives, but this idea was quickly judged ineffective and abandoned. As a consequence, we must return to the fundamental meaning of "democracy," the power of the demos to govern itself. Just as the dictatorship of the proletariat rapidly became the dictatorship over the proletariat, so modern democracy quickly became a power exercised over the demos.

Today one often hears that the population of a democratic country has rights; there are countless speeches on the rights of man. But in reality the people have no power. They neither make the laws nor govern. When have citizens been party to significant decisions in recent times? Whenever a group of citizens come together to examine the policies of an administration, they are always carefully marginalized and pushed to the periphery, as we have seen with today's ecological groups.

If one seeks a democracy, then the demos must become truly sovereign, but not in the manner of the cliché, "The King reigns but does not rule." The sovereign demos must actually govern. Why? Because technique, in its excesses, calls into question the "totality of the human being."[7] As a consequence, there must be a demos regulating itself by its own laws, governing on its own and keeping experts under its thumb. But as Castoriadis has noted, this leads ultimately to an independent autonomous institution and implies the search for an optimum, or at least sufficient, consensus on the essential values that are the goals of governance.[8] Such a consensus must agree on solutions that may feasibly be brought to bear on real problems. This consensus must be obtained in collective, reasonable, unbiased discussions among groups, unlike the discussions commonly held during today's electoral campaigns.

DESIRE AND POSSIBILITY

I am well aware that as soon as the question of democracy is raised in this manner the reader will react by saying that it is impossible and unfeasible. Can the people, as such, really exercise this ultimate power?

Power of this kind implies, above all, an effective control over technicians and the wherewithal to choose either to abandon or utilize techniques. If the people had this power, what might they have decided, for example, regarding the general direction of technical research and its applications? Freed of the fascination of advertising, the people could designate which problems might be important to research. They would have to choose between one technique and another, deciding in terms of cost, ease, danger, and so forth. But all problematic elements would be submitted in a simple fashion to citizen assemblies. In addition, the people would have to sometimes decide on stopping potentially dangerous techniques before they become finalized. These dangers can be located by experimental evidence from an honest evaluation of major risks. Decisions of this kind cannot concern techniques alone, or their research or applications. Behind all techniques are, of necessity, living individuals.

To change institutions so that people might govern would require a real shake-up. The first radical change would require an abolition of the political class in the countries where such a class exists. The political class is composed of professionals, persons who have been elected and make politics their lifetime occupations. They believe that they are in office for life as they capture election after election and end up converting their political responsibilities into a means of amassing a fortune. Obviously, from the moment when politics is professionalized, when a person devotes his or her entire life to the game of politics and exploits its power to personal advantage, democracy disappears. In this context the citizen can accomplish nothing politically without becoming a professional.

In true democracies, however, office holding was always limited, and one could not make a career out of it. This allowed for the continual appearance of fresh blood. In ancient Athens office holders were selected by lot from the entire citizenry. I am convinced that ninety percent of all French citizens would be capable in three months of being as respectable as the deputies we have now. The destruction of the political class should be accompanied by the abolition of the representative system.

But another change is necessary to achieve true democracy. All institutions must possess a maximum of flexibility with a minimum of regulation. In fact, the possibility of democracy is now blocked, not only by the political class, but also by administrative suffocation. Even with a very liberal and non-oppressive system of government, nothing changes if the administration remains in power. Administrations are the real technical dictators in a democracy. Genuine change can occur only by abolishing the regulatory power that silences the voice of the people within the law-making process. A law nowadays is nothing without its "instruction manual" that consists of decrees, stays, and findings, which eventually become the true law. Under these conditions a citizen lacks the power to vote out of office those responsible for misbehavior, the only effective sanction in the representative system, but one that has no effect against the power of entrenched administration.

Moreover, while a citizen may be able to read and understand dozens of legislative texts, he will never find it possible to read the thousands upon thousands of regulatory documents. Not only is the citizen excluded from real power, but, more importantly, he is ignorant of the various rules that prevail in his society concerning what is permitted and what is not. Hence, the need for the abolition of regulatory power and the return of decision making directly to the citizen.

This approach has two important consequences. When a particular administration assumes too much power, when it occupies too great a place, citizens should be able to alter or abolish it. By the same token, if citizens are denied their rights, placed under the guardianship of administration and totally subordinated to it, they cannot have ready access to the state, except through the administration as intermediary. Thus, the state will no longer consider petitions for redress of grievance.

This reflects the single greatest problem facing democracy in our time. If a man wants to be free, he must accept responsibility. If, on the other hand, he wants to be protected in every way, he loses his freedom. The negative concerns we have voiced here (the limitation of administration, etc.) have their counterpoint in the positive need to foster citizen responsibility, the willingness to accept risk and participate in the formation of general responsibility, both at the local level and within larger, collective settings.

If citizens are to experience power, there must be both personal devotion and time set aside for political activity. But here we encounter another obstacle technique creates for democracy. Techniques allow the individual many avenues of escape and diversion, such as motoring about the countryside, which then discourage people from spending long evenings of reflection on large and small-scale community projects or long weekends devoted to working in common on the great political problems. Those who commonly acted on these problems at the beginning of the twentieth century were the true militants in both labor unions and in politics. Today, I would say outright that there are no longer any militants, no longer any union activists; the breed has been destroyed by television and the jalopy.

If democracy is to be possible, it is not enough to remove all obstacles. Man must also assume a responsibility that allows an appeal to a common sense of the world and to the integrity of language. It is, however, quite difficult for citizens to address problems of the nation as a whole. Instead, we must return to small communities, for example, to managing committees of apartments, to workshops, to villages, or to small gatherings of intellectuals. Local autonomy must flourish in all these diverse local domains and do so on a wide scale.

But technique has universalized and homogenized human existence and culture, homogenized ways of living, acting, and working. This universalizing of the economy of Western culture has been accompanied by the cultural and economic collapse of the Third World. Thus, as long as the Third World remains in crisis and poverty, the democratization of our own societies is impossible. Wars in the Third World are now promoted by arms sales which are profitable to commercial and technical development but are injurious and contradictory to democracy; no democracy can exist in a country where entire sectors of the economy depend not on the needs of the local population but on the needs of transnational firms and their various industrial and commercial undertakings.

Democracy will require helping the Third World effectively and without motives for profit and gain. The necessity for the citizen's dedication, mentioned above, needs to be brought to bear to offer unselfish aid to the Third World. Human morality requires that we aid the Third World for its benefit and not ours. This presupposes a thoroughgoing conversion of our techniques and economic system. Indeed, service to the Third World involves unavoidable choices. It is no longer sufficient to furnish the Third World with foodstuffs and medical supplies.[9] Rather we must offer means chosen in relation to practices and natural resources of each region, so that each would be able to meet its needs in its own way. "Business charity" must cease.[10] If we would produce the means of self-development for each Third World country, we must realize that we will no longer be able to move ahead full-throttle on our own technical and industrial growth. Instead, we must pause to equip the Third World free of charge. This will require putting an end to the research and development monster. It means seeking an equality in worldwide living standards and economy that, strictly and rigorously, directs all our technical and economic capacities toward the Third World.

But, here again, we run into a serious obstacle: the susceptibility to corruption of many governments of the Third World is well known.[11] The assistance and implementation of this new policy must not be addressed to those governments alone. The people themselves must be involved directly. Determining whether or not a government is Communist is no longer the problem. Communism means nothing in Third World countries. The possibilities and desires of these countries must be taken as a precondition for dialogue with their people. The problems here are directly linked to ones we have already noted. Accelerated technical progress in developed countries excludes democracy in the rest of the world because it leads to the installation of governments that blindly copy Western models.

IMPOSSIBILITY AND CHOICE

If we are to return to perspectives and activities required in a true democracy, there must be a sweeping change in our present orientation. Two preconditions required for this change must now be faced: competence and responsibility.

How can people who are incompetent make important decisions with regard to technique? Here, of course, ordinary citizens are in exactly the same place as the politicians, who are also perfectly incompetent.

Politicians typically base their technical choices upon spurious reasoning, questionable data, and bogus values (for example, the single-minded fixation on international trade).[12] Faced with choices that are based on aggregates of technical information, we deal with phenomena that are generated by technique itself.[13] What choices and directions shall we take? To affirm that small economic or demographic units are desirable is to acknowledge that such units are better able to judge the appropriateness of technical dangers. But this presupposes the destruction of enormous technical complexes in favor of small-scale operations.[14]

The incompetence of both the people and the politicians raises the dramatic question of the responsibility of technicians. If one does not confront this question head on, there is no hope for realizing democracy. Responsibility is situated on two levels. We must first consider the responsibility of technicians for the projects in which they are involved. Often the projects recommended or carried out by technicians are impossibly expensive and harmful to mankind or to nature. All those who participate in such harmful projects should be held responsible and penalized for any waste or damage. If, for example, a technical operation results in loss of human life, why should the death penalty not be required—a life for a life?

However, the responsibility of the technicians and scientists should not end here. Democracy requires that the people must be correctly informed. If the populace is to make sound decisions, it must have exact and relatively complete information (complicated calculations and experiments are not necessary) regarding the means employed and the dangers that might result. I am thinking here of the fantastic lies told by high ranking scientists on the subject of atomic waste and radioactive mud at Seveso; the misinformation surrounding the consequences of the Chernobyl accident is another case in point. Technicians who participate in such ruinous projects, those who launch forth on technical schemes based upon incomplete or inexact information, must also be held responsible. The higher the technician's rank, the heavier the penalty. This would be perfectly democratic. In ancient Rome, during its truly democratic period, penalties for public officials increased along with their rank.

When I say that the public must be informed, I do not have simplistic solutions in mind; the public must be given information that allows free decisions, not ones based solely upon a menu of options served up by technicians. Scientists and technicians must provide verifiable studies of profit and utility (for example, the TGV, Tres Grand Vitesse train, was notoriously unprofitable). Such studies must seek to reveal, as far as possible, the positive and negative aspects of a proposed operation, for example, its impact studies on human, natural, and sociological environments. The people as a whole, evaluating risks and dangers and goals and advantages, would be left to decide if the game is worth the price.

But even if the demos is aware of all such data, even if it has been informed on every problem on which it is to decide, would this be sufficient? Is it possible to expect the people as a whole to want to choose rationally? Would people be inclined to accept, for example, even a temporary halt to "progress" and to the growth of consumption? Would they be willing, given the importance of the stakes, to accept even a slight lowering of the living standards and comforts of modern life or a certain amount of austerity? At present the masses are manipulated and molded by advertising and appear to be little more than consumers and trend-followers. The truly radical question is: Would the masses remain massified if advertising (the most abominable enterprise of our modern age) were eliminated?

Asking these questions we arrive at what holds the key to true democracy, the first step toward a post-technical democracy. This is the creation of virtue and morality, recognized and experienced by all in our Western world. In proposing this, I am not sidestepping the problem of democracy in its own right. On the contrary. All founders of democracy from the ancient Athenians and Romans to the writers of the eighteenth century and the first revolutionaries of 1789 are unanimous on this point: Democracy is unthinkable without virtue. It is true that virtue has become more difficult in the technical environment because habits and desires are infinitely more numerous, more powerfully anchored in advertising. By the same token, intellectuals today advocate neither ethics nor morality, so consumed are they by the preeminence of desire.

What I mean by morality has no common measure with bourgeois, pseudo-Christian morality or with Victorian morality. It is obvious that such conventional morality is now entirely vain. What then are the conditions for my sense of virtue? I will suggest three. First, we must begin to rely on common sense over and against technical and administrative complexity. Common sense, rare today, effectively locates the weaknesses in technical, scientific, and economic reasoning. Consider, for example, the explosive character of the works of Bernard Charbonneau or Francois Partant, which begin with simple insights from common sense, for example, the recognition that there cannot be infinite and unlimited growth in a finite and limited universe.

Second, virtue of this kind requires a deliberate effort of consciousness. This step is fundamental because the multiple diversions imposed by today's media destroy the ability to focus consciousness.

The effort to achieve true awareness should be carried forth on a number of fronts. In particular, we need an awareness in clearly comprehensible terms of the nature of technical science and of the new order into which we are plunged. To accomplish this we must take every opportunity to flee the cityscape in which most people are now obliged to live. We must also take stock of the real limits of technical science, with its dangers, illusions, and false problems.

Third, we must face the fact that each of us can truly accomplish something, that we can challenge fate and the conditions that now appear to us as grim necessity. We need to nurture a courageous civic sense, an awareness of our true power, all of which should be obvious to anyone who uses the word democracy. After all, what is left of democracy when the average citizen is convinced of his or her powerlessness? An awareness of one's power, however, implies a sense of personal responsibility. One must recognize responsibility for using dangerous technical products such as aerosols and automobiles and understand the way such use defines the direction and development of our society. This sense of responsibility must also contend with the counterfeit discourse of politicians.

Thus, true virtue requires a double awareness of personal power and responsibility that must further be tied to clear thinking, informed by common sense. This capacity for lucidity, this inner light, is already present in each of us. It must occupy a privileged place, nurtured by our schools, so that we might judge the consequences of technique and make right decisions. Clear thinking is an antidote to the all-knowing predictions of the technicians who pretend to serve necessity but who in fact only cast an obscuring veil over true virtue.

A pre-condition of true virtue rests on the creation of a scale of values that addresses the living person. Indeed, there is a crying need for new values. Modern man is lost because values that are truly relevant no longer exist, values based on essential human needs and the affirmation of life.

On the scale of values, the seat of honor should go to wisdom with all that it implies, namely moderation, non-violence, and temperance—measuring each thing wisely and not yielding to technical hubris. Virtue also requires qualities of self-mastery necessary to counter Dionysian flights and the ethos of drugs. Self-mastery corresponds to the ancient sense of modesty (*aidos*) as both a social and political virtue. In the *Laws*, Plato argues that, "Athenian democracy remained admirable so long as a sense of modesty (*aidos*) reigned." In ancient Rome, Cato the Elder offered his countrymen the same advice. If we today want to maintain a chance for life as against the technical order, we must refuse *all excess*: the excess of political extremes magnified by propaganda, the excess of advertising, the refusal of which would reduce greatly the madness of technique and its anti-democratic tendencies that undermine the *sensus communis*. We must refuse the excesses of consumption and its proven economic disutility. We must refuse the excesses inherent in the consumption of energy. We must refuse the excesses of control by the exercise of self-control by each group and each individual. We must refuse the excesses of the profit motive along with the banality of profit-seeking executives and entertainment celebrities. These forms of refusal, consequences of the wisdom of moderation, are fundamental to the existence of true democracy and a practical way to counter the dominance of science and technique.

Although it may seem surprising that I place such importance on purely personal and inner motivation, democracy requires the development of an inner motivation that can rectify the abuses of technoscience. Equally necessary, however, is the understanding that technique can only be mastered by the development of a communal ethics.

CONCLUSION

I know perfectly well that all I have written here appears totally unrealistic and even unthinkable. But if a social group arose, either spontaneously or by deliberate plan, that was sufficiently informed and willing to act with complete awareness as the "guiding force," then my proposal would not seem so unrealistic.[15] But the field is bare: neither the churches, which understand nothing about this question, nor the ecologists, who have been catastrophically politicized, have been able to lay the foundations for such a movement.

There are, however, growing numbers of thinkers in the Western world who are concerned by evidence of an imminent worldwide breakdown in technical, scientific, economic systems.[16] We are beginning to see strong evidence of this breakdown in our daily lives: galloping worldwide inflation, unemployment, and burgeoning military budgets, coupled with a widespread sense of futility as we try to face up to such problems.

The mounting sense of futility often gives rise to obsession or resignation. People are obsessed by the latest thing, the latest gadgets and latest instruments of power—the dominant passions of the modern age.[17] At the same time there is a mood of resignation on all sides. We all know about prominent ecological disasters: the destruction of the ozone layer, rising sulphur emissions, and dangerous chemical pollution.

Faced with these problems, we believe we can do nothing. Thus, our knowledge suggests only passive retreat.

Under these circumstances we are beginning to see an inevitable and growing crisis in our civilization, a breakdown of the financial sector, which will lead to an economic crisis and ultimately a crisis at the very center of technoscience. This crisis is already breaking out on the cultural and moral planes. It will bring about tremendous dislocations, suggesting the need and possibility of returning to the human scale, to individual responsibility, to meaningful personal relationships, all of which will require a degree of austerity. Clearly, the crisis brings great risk. But it also raises the possibility of democracy, a possibility that can be realized by reducing the power now invested in technique.

At present, a proclamation of utopia seems indispensable. Utopian thought is the only way of thinking that can answer the excess of technique and open the way to essential hope. Utopianism does not necessarily mean unrealistic fantasizing. Instead, utopias point toward alternative social models that might be implemented.

Each of the three ages I have discussed here generated a magical interpretation of the universe. With the rise of the social order, for example, mythical thinking became the predominant mode of understanding the cosmos, to be superseded at a later point by the development of ideological accounts of the world. In the present crisis, we find a profound rupture in the social imagination and its symbols. What attraction can be found any longer in progress, well-being, growth, and the other values politicians still throw at us? These values are without meaning to our society and its history as well as to the technical order. By contrast, technology and utopian thought share two fundamental characteristics. While technique destroys the specificity of place, utopian thought explores the placeless. Further, technique propels us toward a meaningless future, while utopian thought expresses an openness to a future of possible meanings that may be reborn.

Defining utopia this way, we see that our models are not to be found in the classic utopias of the sixteenth to the eighteenth centuries. We need a new dimension for this kind of thinking. Technique has devalued myth, including the myth of the rights of man, which the French people so proudly celebrated during the bicentennial of the French Revolution but denied in their actions. Technique has absorbed the sacred and has, in fact, become the sacred. Faced with these seemingly insurmountable realities, we must invent another mode of being: an iconoclastic democracy capable of desacralizing technique. In this light, utopia offers itself as a reconciliation of the transcendent and immanent in an expression of thought sought explicitly or implicitly by everyone. From this would come a challenge to modern man the consumer, who is not satisfied with conspicuous consumption, and to modern man the spectator, who often yearns to be an actor. The challenge is for such a person to accept the responsibilities and options that now confront him. But this can only happen if a new utopian model offers itself and is brought before his eyes.

Bordeaux, France
[Translated by David Lovekin and Michael Johnson,
Hastings College]

NOTES

1. Jacques Ellul, *What I Believe*, trans. Geoffrey W. Bromiley (Grand Rapids,Mich.: Eerdmans, 1989), pp. 104-140.
2. See Jacques Ellul, *The Political Illusion*, trans. Konrad Kellen (New York: Vintage Books, 1972), p. 3, note 1; *L'Illusion politique* (Paris: Lafont, 1965), p. 11, note 1. [Translators' note: Ellul distinguishes between *la politique* and *le politique* in the following way: *le politique* refers to the general goals of political acitivity, considerations for example of the type of good life a political system should serve and sustain. *La politique* refers to the machinery of the political system.]
3. Compare this to my *Technological System*, trans. Joachim Neugroschel (New York: Continuum, 1980).
4. See for example the excellent book by Nöel Mamère, *La Dictature de la'audimat*, published by Le Découverte, in which he demonstrates that if any high quality presentation does not immediately catch the public's eye, it is quickly withdrawn according to the decree of the "audimat," a collection of instruments that measure audience response.
5. I have already shown this in *The Technological Bluff*, trans. Geoffrey W. Bromiley (Grand Rapids, Mich.: Eerdmans, 1990). The mindless making and consuming of gadgets has nothing to do with efficiency.
6. M. Duverger, *La Tentation de l'impuissance* (Paris, 1988).
7. Let there be no misunderstanding of the term "human being." The word "being" is a verb and "human" is a noun and not an adjective.
8. It is clear, for example, that "profit," the maximum gain, and "efficiency" are not principles compatible with the pursuit of efficiency as an ultimate end.
9. De Ravignan has shown that food aid is often destructive to the populations that receive it. For example, furnishing free rice destroys small-scale local production of rice. Peasants in that locality can no longer sell their harvest when Western rice is distributed free of charge.
10. This concept comes from B. Kouchner.
11. We encounter here a phenomenon that deserves closer study: the role played by corruption. This affliction appears to affect all governments. It would be interesting, for example, to count all major cases of corruption uncovered during the past ten years in the United States, France, Germany, and Japan. Our governments are themselves afflicted by a disease that, for a long time, seemed to beset only Third World politicians.
12. Consider, for example, the deception involved as people invoke the idea of freedom of expression as grounds for the privatization of television. Such privatization has made television a complete slave to advertising firms.
13. Compare this with my book, *The Technological Bluff*, pp. 33-120, where I discuss

unpredictability, ambivalence, long term planning, and the like.

14. Why could not electricity be produced by windmills or water driven turbines on small tributaries in place of the monopoly of the gigantic E.D.F.?

15. I am thinking about socialism, which in the nineteenth century could exist as a well-rooted and not mythical doctrine because it had its own supporters in the workers.

16. The sense of political breakdown in the West becomes evident in, among other things, the growing number of non-voters. Forty percent of eligible voters abstained in the European elections of 1989.

17. Two instances of public reaction to proposed policies offer an illustration of this point. In 1980, as a part of its anti-noise campaign, the French government decided to prohibit the use of unmuffled motorcycles and other "two-wheelers." In response, the Place de la Concorde was taken over by thousands of bikers. After a series of protest marches the government gave in. Similarly, in 1988 a bill prohibiting the manufacture of touring cars that could exceed 100 miles per hour was withdrawn after enormous demonstrations to show that the public is determined to drive at high speeds.

ROBERT L. FROST

MECHANICAL DREAMS: DEMOCRACY AND TECHNOLOGICAL DISCOURSE IN TWENTIETH-CENTURY FRANCE

Someone examining the technological artifacts of France over the past century might be stunned by their epic scale and profound evolution. Perusing artifacts at each end of the past century, the Eiffel Tower and the plutonium breeder reactor at Creys-Malville, can elucidate these aspects. While their large scales certainly evoke strong reactions, their respective meanings are vastly different. The Eiffel Tower reflected republican engineering, with each equally-sized member structured rationally and functioning harmoniously as the eminently public monument reaches skyward. By contrast, the breeder communicates its technocratic-authoritarian style by several steel strata of security shielding, which house plutonium, the deadly namesake of the ancient Greek underworld, far from the public. The public can freely mingle at the Eiffel Tower unhampered by police or guards.[1] The proprietors of the breeder grudgingly welcome the public with a solar-heated visitors center (placed outside the security perimeter), replete with heroic, pressed-plastic photos of experts designing, building, and operating a seminal symbol of energy engineers' expertise.

An effort to decode the meanings of Eiffel Tower and Creys-Malville would have to proceed from a recognition that the two artifacts are social and cultural constructs as well as mechanical structures. The Eiffel Tower, a French revolutionary centennial structure, reinvented the "republic of virtue" to become a republic of functionalist reason. The monument, though constructed of brittle iron, used its thousands of equal (yet hierarchically-arranged) members, bound by pivotable rivets, to flex and adjust to changes in the weather. Open and accessible, hierarchically egalitarian, flexible, and speaking to an aesthetic of machine progress, the Eiffel Tower stood mockingly as a critique of traditionalist anti-republicans. By implication the latter were closed (politically, socially, and intellectually), dysfunctionally inegalitarian, rigid, and opposed to human-mechanical progress. In contrast, Creys-Malville, the leading icon of the French technocracy, symbolizes the consolidation of a new elite. The product of public funds, private suppliers, and publicly-educated experts, Creys-Malville begs the public to ignore the "technicalities" of progress and to accept the bounty of plentiful power. The plant's scale, rather than inspiring awe at a technical-democratic aesthetic, intimidates the hapless visitor. It artifactually argues that only experts are capable of making technical-industrial decisions, and that politics and popular intervention are dysfunctional. A demonstrator was shot in 1977 as he tried to demand more open public debate on energy issues. To the builders of the plant, the incident was a police issue, not one of technological choices. As symbols, the Eiffel Tower and Creys-Malville not only reflected the political values of emerging elites, they *created* and *reconstituted* them. The former empowered as the latter disempowered.

These two technological-cultural signs, between which the heart of France spatially lies and the past century temporally extends, shall serve a dual purpose for this article. They will function as emblems for a theoretical discussion about the making and workings of technological culture, and they will help to elucidate the evolution of French technological culture, particularly since the Second World War. This article will first develop theoretical, methodological, and explanatory links between technology and culture in a broad sense, particularly as they pertain to technical, political-economic, and cultural-symbolic change. It will then examine the postwar French economic miracle, using the notion of technology as heroic narrative to explain the French preoccupation with technological display, and how a new discourse about Progress forestalled genuine political debate. We shall close by examining a political nexus in which debates over merely the form of ownership—not the products, practices, or style—of major firms has become the substitute for real contention over the shape of France's workplaces, worklife, products, and distribution of economic and political power. This paper argues that popular ideology has gradually reified technology, so that technological images have overshadowed their function. Public debate about the ends of technological change has run up against a premature closure because reference to expert-dominated technological sign systems has helped to shift decisions about technology and political economy out of the public forum in contemporary France.

Recent scholarship in the history of technology sheds new light on the connections between technology and culture. Earlier approaches within the history of technology tended to focus upon heroic inventors and necessary or critical inventions. From this point of view, necessity is the mother of invention and the history of technology is a linear evolution of single best solutions, as Robert Heilbroner termed it, "[the] technical conquest of nature that follows one and only one grand avenue of advance."[2] This internalist and technologically determinist approach makes for societies as spectators and inventors as heroic discoverers.[3]

More sophisticated analyses have investigated the invention and diffusion processes. Arnold Pacey and Jeffrey Meikle, respectively, show how broad cultural and personal styles and tastes—from cultural predilections for scale to personal æsthetics of mathematical or design elegance—drive the inventive process and end up being built into technological artifacts.[4] Parsonians and neo-classical economic historians link a dual, almost transcendent co-functionality to inventions and businesses. David Landes and Alfred Chandler well portray the magic mix of astute design, bold vision, and organizational acumen behind successful technologies. They nonetheless imply that technical efficiency (however defined) is virtually identical with economic efficiency (however defined), provided lower forms of intervention, such as politics and social demands, do not intervene.[5] Landes and Chandler argue that scale, rationality, and scientific business methods have interactively been in a constant state of evolution toward higher forms along a trajectory of single-best technico-economic solutions. These authors do, however, offer a valuable conceptual parrying at the problem of how societies and technologies shape each other by offering a range of useful techniques and theoretical frameworks.[6]

Thomas Hughes's and John Law's concentration on the growth and development of technological systems—broadly defined to include individuals, social groups, and the logics intrinsic to specific technical systems—offers a far more promising tack.[7] These authors grant equal status to technical and social forces in shaping technological systems, stressing the complex interaction between technical logics and and social forces.[8] Hughes's conception of momentum (again, a social-technical construct) offers a way to understand how past choices impinge upon later options,[9] and his notion of "reverse salients" gives valuable insight into dysfunctionalities among elements within and contiguous to social-technical systems. Systems analysis requires that system boundaries be precisely defined, yet the new history of technology denies the utility of system boundaries—not only did a British jet fighter of the 1960s have links of differing strengths with various producer-promoters, it had them as well with state agencies and potential users.[10] This sort of systems approach owes much of its power to the flexible boundaries it allows actors themselves to draw around putative systems, yet in using such loose boundaries, this tack loses much of its heuristic appeal.[11]

Hughes's conception of momentum shows how choices within technical systems at one time tend to prestructure later options, and how social dynamics and structures (one is tempted to assert *subcultures*) associated with specific systems bequeath legacies. The concept of momentum has within it, however, a subtle conception of movement upon a unilinear vector, and one is left musing about whether, for example, power system reliabilities could better have been served by a larger number of smaller facilities, especially after returns to scale level off in mature technologies.[12] Hughes's notion of style, which for him seems to be national, can easily be seen as subcultural as well.[13] At that level, subcultural technological style can become analogous to the cultures specific to individual firms or technocracies.

More radical approaches to the society-technology issue appeared in 1980, 1983, and 1984 with, respectively, the publication of a special issue of *Culture technique* on domestic technologies, Ruth Schwartz Cowan's *More Work for Mother*, and David Noble's *Forces of Production*.[14] These authors adopt a largely socially determinist notion of technological change, in which artifacts are relevant only within social contexts and where those upon whom new technologies are foisted remain passive or mute. To use literary language, from this perspective, the technological text is meaningless outside of its social context. Furthermore, an overdetermining context compels inventors, innovators, and diffusers consciously or subconsciously to build often oppressive social relations into artifacts. In *Culture technique's* and Cowan's view, the relevant context is gender, and in Noble's, it is social class. Context often limits imagination, yet for Noble, innovators consciously chose technologies which reproduced structures of class domination. Implicitly or explicitly, this genre of scholarship built upon Langdon Winner's highly innovative work.[15]

One may question the uncompromising social determinism of this work. Though it can be shown that some better technologies failed because of their rejection by dominant social interests, this does not prove that technical performance will necessarily always be subordinated or controlled by inventors or elites.[16]

Though Noble allows for choices and he may be correct about the cynical classism of machine tool designers in their quest to deskill labor, one wonders if all inventors are so prescient or powerful. The omniscient rational actor of the Parsonians appears here in contrasted colors. A flat unilinearity can re-enter this debate through a social determinist back door. Nonetheless, these approaches often elegantly open the technological black box, arguing that the internal technical characteristics of technological artifacts are set in some detail by external forces. Equally important, they show how historical actors invent idealized futures and plot social routes toward them.

Sociologists and anthropologists have recently helped to clarify several issues in the debate about the interrelationships between society and technology. Students of the social construction of science have broken through the common intellectual division between a natural science of hard facts and the human sciences of interpretation.[17] Using a notion of "interpretive flexibility," Wiebe Bijker argues that in practical terms, artifactual "facts" are actually only sets of social meanings.[18] What is materially behind the image is irrelevant because it cannot be known. Thus, the modes for constructing meaning and the social groups which attribute meanings offer the integral avenues to look at the techno-social artifacts.[19] Bijker focuses on "relevant social groups," whose interests are built into artifacts and later inferred by historians, as the key to discern the social meanings and impacts of artifacts. The problem of defining social interests remains paramount.[20] Bijker nonetheless crucially offers a way to understand how the social meaning and physical shape of artifacts become stabilized by defining a process of closure, in which debate over the meanings of artifacts essentially ceases and designs and social roles solidify.[21] Nonetheless, by divorcing social contexts from technical attributes, this approach tends to be overly social-determinist, rendering the actual designs and physical attributes of technology as simple, flat outcomes of social forces and leaving little latitude for recognizing that certain laws of physics might also have considerable impact—it begins perilously to reclose the black box. In addition, Bijker stresses contention over meanings, but he does not allow for negotiated redesigns of artifacts, nor does he grapple with the ways that meanings are constructed, except through the methods of organizational sociology.

Bruno Latour and Madeleine Akrich have developed ways to understand how both the meanings and material content of technologies can be adjusted in order to implant both artifacts and innovators. They also indicate how closures are at best provisional and subject to renegotiation when the social and technical networks supporting them weaken.[22] Using sociologists' resource mobilization framework, they show how all parties to an innovation—inventors, recipients, and artifacts—adjust artifacts, meanings, and social networks to make new technologies and contexts socially credible and powerful.[23] A process of implicit negotiation (*translation* in Latour's terminology) modifies artifacts, meanings, and contexts. Innovation and diffusion therefore become a single, dialogic socio-technical process, indifferently redesigning physical characteristics and reshaping social coalitions as power and resourcefulness demand. This approach represents an important development, yet it seems to presume that the social actors somehow objectively know and consciously act upon all social and technical variables. Latour tends to ignore the larger social, ideological, and political structures which shape and set the parameters of decision making.

Cultural anthropologists have long argued that the construction of meanings is an imaginative, yet culturally structured, process and that those who create meanings need not be aware of all aspects of their context. They offer ways to view artifacts as complex texts susceptible to reading and decoding, but with the additional twist that until closure, designs and meanings are constantly shifting. Bryan Pfaffenberger, a cultural anthropologist, has recently presented a first foray into this set of issues.[24] He indicates how people relate to new technologies by showing how, intentionally or otherwise, signs—syntheses of signifiers and signifieds, images or words and artifacts—attain meanings, and how people act upon them. Within this framework, meanings can be shifted and reconstructed according to changes in the real or perceived social and material context. Because such processes of adjustment and reconstitution must draw from culturally available meanings, they most often serve to mirror, and, over time, reproduce the existing social-cultural order. By this argument, there can be no genuinely revolutionary technologies. Nonetheless, this approach does not suggest a clear empirical technique for linking people who are acting within specific cultural frameworks to real objects and cultural signs.[25] This approach seems to assume that the material characteristics of devices remain rather fixed and that adjustments of meaning do not entail adjustments of objects beyond what is necessary to meet socio-cultural demands.[26] Pfaffenberger describes the process of meaning adjustment and reconstitution as "a rather pathetic quest for self-validation in terms laid down by the dominant culture,"[27] which offers only a limited foray into understanding how images *and* artifacts are constructed and reconstituted.

Similar problems emerge in studying more literary and structuralist explications of sign systems. Their approaches do, however, suggest ways that we can see the link between social meanings and artifacts because the latter are, as "texts," susceptible to being read and deconstructed. Roland Barthes offers ways to analyze the apolitical and anti-conceptual ways that popular ideology is promulgated,[28] and how meanings are impoverished and manipulated while perhaps preserving the physical integrity of the artifact.[29] For Barthes, myths—linguistic and symbolic constructs which replace the real objects they once represented—can form the basis for perceptions upon which people act, for example, in pursuing imperialism, purchasing consumer goods, or (for our purposes) inventing new technologies.[30] Nonetheless, structuralists, from Lévi-Strauss to Foucault and Barthes, render the recipients of images passive and cannot explain actions *in response* to myths. Were historians of technology to adopt this pure form of structuralism, we would not only destroy the heroic inventor of the antiquarians but would also arrive at a history without conscious actors, not unlike that of Braudel. These approaches do offer ways to understand sign systems themselves and (with some major modifications) the available fragments out of which people construct often disordered meanings for objects. For example, as a sign, a flag has no more or no less meaning or significance than a nuclear plant or playground incident; it only occupies a different place in a constructed world-view assimilated by a passive receiver. Nonetheless, we have no way to understand how people might personally link disparate signs—how or whether a flag-sign might link to an automobile-sign or a vacation-sign.

We can look to conceptions of narrative within both literary criticism and cognitive psychology as ways to link symbols and actions, and to understand how people might assimilate and adjust available technological signs.[31] Some cognitive psychologists argue that individuals construct narratives in order to understand change. Stories are constructed to organize signs, events, and processes—indeed, the entire panoply of perception.[32] Historians have recently begun to recognize the value of the narrative approach as a tool of analysis.[33] Along with Bijker but from entirely different perspectives, students of narrative identify moments of closure, the point at which contention over meanings ends and new signs become stable.[34]

Those who stress the significance of narratives most often offer them as tools for retrospective analysis or for literary criticism, viewing narrative construction as a way of serially interpreting the past or of interpreting a text. Historians trying to understand the changing character and meaning of artifacts or the intentions of historical actors need a conception of *prospective* narrative. Prospective narrative, or scripting, is thus the process through which historical actors visualize futures and invent scenarios to arrive at those futures.[35] Just as consciousness is constructed by a subjective and half-conscious ordering of the terrain of available signs, conscious actions (be they political, inventive, or social) are directed to a reordering of signs and meanings—toward closures—by scripting future histories. Scripts are, of course, constructed by each individual with implicit weightings arbitrarily assigned to constituent parts. In addition, Madeleine Akrich has correctly argued that inventors prescribe (or pre-script) the meanings of artifacts and socio-technical processes within designs themselves.[36] Recipients must therefore negotiate their own future scripts against those pre-scripted in specific artifacts. This discursive process affects both the meaning of the artifact and the self-identity of the perceiver. The flexibility of an artifact's built-in pre-scription helps determine its ability to adapt to different and changing circumstances. A liquid propane gas range has far less cultural or material adaptability than a cast iron stove; the former demands a specific fuel and social environment, the latter is indifferent to fuels and requires fewer social preconditions. Non-negotiable pre-scriptions or unilateral inventions are less apt to be socially assimilated because they are inherently less adaptable.

Scenarios are important for us as integrators of signs and media for meanings, and when specified in fine enough grain and viewed from outside, as definitions of frontiers between structural and individualist determinants. Indeed, on this latter point, if people are to retain a sense of power and membership within their social context, they must constantly re-invent scenarios to take account of their shifting perceptions of what is possible and what is structurally forestalled or inevitable. Similarly if promulgators of signs (political parties, inventors, advertising agencies, etc.) are to remain viable, they must also adjust signifiers to meet the changing priorities of recipients.[37] Thus the process of adjustment occurs on both sides of sign systems, and it encompasses both meanings and contents. As such it is dialogic. Just as meanings can be adjusted or reconstituted to take account of new circumstances or the intrinsic character of artifacts, so the design and utilization of artifacts can be modified: airfoils could be added to cars and computers could be used to cast astrological charts. By observing and inferring prospective narrative and adjustment strategies, we can therefore start to transcend some of the thorniest historiographic tangles—between structures and wills, and among artifacts, meanings, and actions.

What makes prospective narratives valuable for historians are their commonality and contrasts—the linkages which arise primarily from common experience, and the differences which lend the definition. More significantly, the successful promulgation of signs (as well as the ultimate hegemony of specific sign-systems) can be seen as supplying a society with a set of meanings which explain seemingly disparate phenomena. The successful appeal of a party program, advertising campaign, or new invention can to an extent be ascribed to success in inducing individuals to identify a meaningful diagnosis and prognosis in a set of propaganda materials. A propensity for people to take certain sets of sign symbols seriously hinges upon the perceived legitimacy of the promulgator, upon the latter's successful references to culturally available meanings, and, if necessary, upon an appropriate adjustment of an artifact.[38] By this avenue, we can begin to conceive of communities which are constructed out of sets of commonly recognized meanings. Safe, clean, and cheap nuclear power made for a nation of warm and contented power consumers; expensive potential bombs in the countryside made for communities of anti-nuclear activists. One scripted a future of a convenient, labor-saved lifestyles; the other envisioned an Orwellian nightmare of "electro-fascism."[39]

In sum, earlier historians of technology who analyze the inside of the black box referred to an engineering conception of functionality or an economic sense of cost and benefit characteristics. Both of these approaches (and a combination of them[40]) tended to be reductionist. At the same time, few more recent authors attempt to examine concretely how ostensibly socially-determined technical characteristics of artifacts serve to reproduce, extend, or undermine existing social relations.[41] Historians of technology are becoming more aware of the symbolic content of technologies, and how the imagery of functionalism and utility may be more important than their reality. If we are to discuss technology as a cultural dynamic and social process, we can enrich our analyses by showing how people imagine, interpret, and interact with artifacts. Machines and artifacts are both evocative, to use Sherry Turkle's term,[42] yet the meanings and attributes evoked are often far from transparent or functional in the usual sense. Machines and the social meanings of them may serve psychological, social, and cultural ends which can exist quite apart from more mundane mechanical purposes. Historians of technology have long told us how artifacts can help beget new social formations. If, as Lynn Hunt argues, "Symbolic practices [can call a] new political class into existence,"[43] technological symbols can create new forms of social identity and organization. While cultural analysts of technology have richly shown how promulgators of technology seek to imbue specific social groups or societies at large with determinate sets of meanings,[44] few have grappled with how and to what degree such images are either assimilated or acted upon by putative target groups. We are often left to assume that images are impressed into a social *tabula rasa*. A theory of technological narratives allows historians to address the ways that promulgators and recipients might have negotiated over the material content and meanings of technological artifacts.

Historians and philosophers of technology might use this framework to study the relationship between society and technology, particularly to define the presuppositions which people brought into their relations with technology. While it might appear that this method re-closes the black box (or worse yet, shows little interest in revealing its contents), it allows the relevant actors themselves to open the box—to define which issues are relevant. For example, Jean Baudrillard has argued that in domestic technologies, the message of functionalism in appearance and arrangement often belies serious lack of functionality.[45] To borrow from Barthes, mythologies of high technology and expertise may well have served to keep the black box closed, even when it was empty. One only has to view the objects of the current minimalist, high-tech style to understand the contradiction between the appearance and reality of function. One might also argue that black boxes have at times been consciously kept closed, particularly those which contain representations of skills, such as computers and special purpose machine tools. From a technical standpoint, some boxes remained closed because "closure of meaning" (in Bijker's sense) had already occurred, as with American cars between the mid-1930s and the early 1970s. Some may have remained closed because the subculture which controlled them did not consider some issues as even existing—hence the automotive carburetor remained untouched with respect to fuel economy until 1973, when signs of (as it were) greedy oil blackmailers and gas lines forced a redefinition and redesign of the carburetor.

Perhaps the most appropriate way to examine the ways to use this method is to return to our example of the Eiffel Tower and analyze its position as the first step toward the closure of technological discourse in France. The Eiffel Tower, an icon designed for the 1889 centennial of the Great French Revolution, represented social-technological republicanism at a time when French progressives were battling for a secular republic. Opposed to conservative, Catholic, and family-oriented business, the tower bespoke a critique of the old order. As such, it was one of the first images in a long series deployed by an emerging social technocratic milieu, and its very existence imagematically attacked traditionalists. Nonetheless, while it represented a goal for the people who wanted to get France out of a miasma of technological, social, and economic stasis, it obviously remained mute on the way to achieve a modern and rational republican régime. The iron and structural concerns of the Eiffel Tower were turned to rapid industrial growth in the years up to 1914 and symbolically began to promise a world of plenty once the fruits of prosperity were better distributed. World War One inverted the imagery—technological "progress" had provided the means for mass destruction—but ironically, many imagined that French industry and technology (not indigenous perseverance or American troops) had won the war. The static tower had become the dynamic and victorious assault tank. As the war wound down, many envisioned a postwar world in which the output of mass production industry would serve broad and prosperous popular markets.

Images in the 1920s, from those of domestic appliances to power plants, underlined the gap between visions of the future and the feeble means available to get there, but by that time businessmen affirming the value of a technology-driven growth economy had elicited the support of trade union reformers and socialists such as Jean-Louis Breton and Hyacinthe Dubreuil in their battle of images.[46] Leaving the Communist Party in an isolated adoration of the Soviet Union and criticizing conservative business for its production-inhibiting "malthusianism," leaders of the General Confederation of Labor (CGT) and elements of the Socialist Party allied with modernist businessmen to promulgate an image of symmetric technologized environments: the rationalization and

Taylorization of factories, and a modernization of homemakers' domestic tasks.[47] Technology as liberator from material want and physical drudgery became a powerful image for this emerging modernist coalition, but the symbols remained hollow, static, and without concrete referents. Indeed, those who sought to import Ford's vision of mass production for mass markets had no credible scenario for implementing their visions, as political power and economic policy control remained in the hands of a reluctant Center-Right. Industrial rationalization (which included a range of activities, from reorganizing the labor process to developing production-cost analysis models and notions of material flows within factories) was limited to a few leading-edge industries such as electric power and automobiles. The politics of rationalization was discussed only within small circles of experts.[48]

Interwar technological imagery ironically succeeded—popular visions of a heroic, technologized future became firmly implanted and formed the basis for later political aspirations—despite minimal implementation of the new techniques and technologies. In part, the failure can be ascribed to a lack of credible scenarios to meet the new goals. Most proprietors were loath to risk the capital-intensive investments in new machinery that innovation in manufacturing required, particularly because they sensed that their markets were too small to warrant the effort.[49] In addition, most entrepreneurs still affirmed the traditional French production model of small batch production of luxury products executed by skilled artisans and sold at high prices, and indeed, many gentleman entrepreneurs were reluctant to believe that quality could be mass-produced. Artisanal production had worked for their forefathers and few saw reason to change. Finally, the Fordist model seemed to imply a shift toward managerial capitalism, thereby removing "divine right" owners from control of their firms. A shortage of technocratic managerial personnel, whether business experts or engineers, further compounded the owners' reluctance.[50] In sum, the chasm between heroic visions and the banal quotidian underlined the fact that productivists—those seeking to maximize output through rationalization and other efficiency measures—could offer no credible scenario to reach the future.

The history of the Salons des Arts Ménagers, or annual home fairs, echoes the problems of industrial rationalization. The salons, held in Paris from 1923 until 1980, enjoyed massive attendance in the interwar era—almost a half-million in the late 1930s—yet few families bought appliances other than clothes irons or radios. Reasons for the gap abounded. Designs rarely stabilized in the interwar era because of a broad array of energy sources (coal, charcoal briquets, kerosene, gasoline, natural gas, and electricity) and a large number of small manufacturers who remained wedded to artisanal techniques.[51] Design closure also remained elusive because manufacturers, lacking market research capabilities, often imagined what their potential markets wanted and left little room for negotiation with consumers.[52] As a result, they often simply imported American images of the users of appliances.[53] Defining domestic consumerism meant inventing new social roles for women, and wives and husbands themselves were not yet sure about what shape the new roles would take.[54] A new, class-blind definition of femininity, of women liberated from household drudgery by machines, found little resonance either within the elite (which dreamt of servants) or the working class (which was too poor to afford the new machines). The budgets of most families largely precluded items deemed to be luxuries, and even the most primitive sense of accounting would show that pricey home appliances could never really be amortized.[55] Middle class women were beginning to shift toward new conceptions of public life based on paid labor and education, and to the ad-man's romanticization of domesticity they responded with a hard-headed practicality.[56] Finally, urbanites faced a dire shortage of quality housing, and presumably, additional domestic expenditures would first go toward housing itself.[57] Hence, neither manufacturers nor putative users of domestic technologies could agree on the social meaning of appliances or housewives. There could be no credible path to women or families liberated by technology without the closure of appliances' physical features or social meanings. In the blocked society of the interwar era, few offered credible scenarios to reach an idyllic future.

Thus, the interwar era in France saw only a proposal for marriage between consumerism and productivism, never its consummation. Trade shows, expositions, and managerial revolution literature remained signifiers without material referents, yet members of the modernist coalition also promulgated them as critiques. By the 1930s, however, the Communists began scripting a mass-production/mass-consumption scenario based on the heroic Soviet successes under Stalin. Until 1935, that prospective story was, of course, an agenda for social revolution. For the French Communist Party (PCF), in the USSR, fanatically productive workers joyfully worked under "comrade managers" to build a workers' material paradise, replete with massive steel mills and hydroelectric projects. In the interwar era, the PCF thus also began to use its version of heroic technology as a criticism of capitalism, but once the PCF consented to support the Center-Left Popular Front coalition in 1936, its images again lacked scripts.

Indeed, the failure of the Popular Front, despite its traditional appeal to social solidarity and its occasional references to technology-as-social solution,[58] can be ascribed to its lack of any coherent program offering more than piecemeal reform. Based in sentiment more than programs and promulgating images of technological prosperity without content or coherent means, the Popular Front could do little to put France's economy back together. Maintaining a coalition against fascism demanded nearly the full attention of the Popular Frontists. State-financed heroic dam projects in the Rhône basin—reminiscent of TVA—required years of construction time to benefit consumers, and their employment effects were minimal. The régime wished to increase the size of the economic pie as an alternative to warring over its division, but failing such, it fell amid class contention. The Center-Right managed to elicit high levels of productivity from 1938 to the defeat of 1940, but they were largely based upon a yawning fear of impending invasion.[59]

France's defeat in 1940 was broadly viewed not only as a military affair, but as an indictment of an entire class and social system. As a result, the traditional élite was forced into retreat after the war. A leading wartime resistance study group wrote,

The "armistice with honor" posed the problem clearly to our people because it had been preceded by the failure of almost the entire ruling elite. After 150 years of rule, our bourgeoisie, who controlled the army, the political system, and industry admitted its sterility and could conceive of national renewal only in defeat.[60]

In addition, the Depression and war had clearly left France in dire material straits; she was woefully short of capital, labor, and resources. Technocratic reformers therefore began to write the scenario for a French renaissance. On the advice of Stalin, the PCF rejected a revolutionary strategy and became part of the modernizing coalition—at least for a short time.[61] The political preconditions for a cooperative relationship between business and labor were set, and the modernization scenario emerged as a combination of nationalization and planning. The three unlikely personalities heading these efforts were General Charles de Gaulle, the virulently nationalist president, Jean Monnet, the technocratic father of the plan, and Marcel Paul, the Communist Minister of Industry.

Despite very real political differences within the modernization coalition, there was a consensus on a number of material issues. All recognized the dire capital shortage, and that borrowed or granted capital would not suffice. Capital had to be accumulated, not unlike the "socialist accumulation" overseen by Stalin, and this accumulation demanded a vast shift of income uses from consumption to investment in an already impoverished population. The producer-consumer imagery of the interwar period had to be untied, and consumerism had to be a dream deferred. Success in this venture ostensibly depended upon the willingness of the PCF, now the dominant influence in the General Confederation of Labor (Confédération Générale du Travail, CGT), to pursue a "battle of production" at long hours and low pay, and under the command of technocratic managers. In return for its efforts the CGT was offered a vague participatory role in management and planning organisms. The PCF began to script a future in terms not unlike the socialist-realist art of the Soviet 1930s, replete with heroic workers, dogged party and union leaders, and "patriotic bosses" building the New France.[62]

The modernization coalition also developed a consensus on the style and imagery associated with the modernization, and indeed, to an extent, on the scenario for "progress." Reacting to the traditionally small scale and decentralized character of French industry, modernists envisioned a national technical apparatus dominated by large, centralized systems, owned by large, centralized firms, all integrated by "indicative planning"—a system of capital and resource allocations for the private firms that voluntarily followed state-set industrial guidelines. While there were certain disagreements about which enterprises were to be public or private, notions of scale and centralization reached the colossal proportions of a Barthesian myth.[63] Reflecting the rationalist-positivist traditions of French engineering and political discourse, modernizers also tended to place high value upon mathematical demonstrations, often (as in the case of planning) appreciating the æsthetics of "ideal" resource allocations, for which the actual resources and the cash to buy them were often unavailable.[64]

Communists thus affirmed a rather solidaristic and heroic vision of industrial renovation, while the vague Left-Center bloc of Socialists, Christian Democrats, leading-edge businessmen, progressive Catholics, and state managers supported a more managerial and technocratic one. Ironically, though the Communists retained an explicitly political language in discussing technological renovation, both parts of the modernizing coalition tended to use politics to attain goals they judged would launch France into a historical stage beyond politics, and they were to use a rhetoric of linear progress to do so.

Building on the "objective" character of the elements needed for modernization, both visions of a technological future tended to reify and depoliticize technology. Dams, mills, and rail networks of an epic scale and with largely uncontested technological characteristics were to be emblematic of a new France in the making. After all, the Communists could look to the USSR and see steel mills and power systems nearly identical to those in the USA, to which the other wing of the coalition made reference. Even on the question of ownership, the majorities in favor of nationalizing coal, rail, banking, insurance, electrical power, and Renault were all quite lopsided. A long debate on the nationalization of electricity indicated less direct opposition than an opportunity for neo-liberals to take rhetorical stands against what they saw as the creeping power of the state. Indeed, most of the opposition voted for passage of the bill to nationalize electricity.[65] Modernists thus enjoyed near-total unanimity on the only vaguely "political" aspect of the modernization effort, that of nationalizing a number of key industrial sectors, including coal, oil exploration, gas production and distribution, investment banking, insurance, rail, public transport, and Renault. Though it of course remained unspoken, many sensed that even if France had not really won a major war in almost two centuries, her techno-intellectual skills still warranted her a place in the sun, a sun which was to become nuclear. Indeed, the massive modernization projects of the late 1940s and early 1950s were to be artifacts of technological display, symbols to a dubious world of France's greatness.[66] The symbols had important political functions, for they helped to recruit erstwhile conservative businessmen and radical labor activists into the modernization coalition, and to persuade a rather localist society that technocrats in Paris could be the guarantors of prosperity.

The contrasts among the different prospective narratives of the path toward France's future reflected subtle differences, however. Neo-liberals envisioned the tasks ahead as necessary pains on the way to re-establishing a healthy, prosperous capitalism with less dramatic income gaps.[67] Technocratic socialists saw the modernization as a first step to a white-collar, expert-managed socialism—a socialist version of Thorstein Veblen's vision—in which more technically efficient state firms would defeat private firms in the market, and workers and outsiders would defer to expertise.[68] Beneath thin socialist-realist imagery, the Communists' positions were more ambiguous. While it seemed that they were embracing the modernization consensus on a permanent basis, particularly because they had won an institutionalized position within the industrial régime for the CGT (over which they won political control in 1946), they seemed to cooperate only as a way to gain power within the new institutional frameworks.[69] The implication of all of these strategies was clear, however, that a dogged campaign for greater production in the near term was ultimately to give the material basis for a transformation of everyday life in the form of more consumer goods. A technological liberation from drudgery was to follow from a similar liberation from penury.

The Cold War eventually doused the enthusiasm of the modernization coalition. By late 1947, the Communists were in the opposition and the Right was on the path to renewed power, which it would gain in 1950.

Though the coalition died at the top of politics, it continued among social networks within the CGT, state sector, and modernist businesses. In Thomas Hughes's terms, the coalition and its projects had already developed considerable momentum, supported by Marshall Plan funds. The Communists in particular were placed in a deeply contradictory position. While the party doctrinally argued that nationalizations and planning were only ways of rebuilding capitalism, and that the program of deferred consumption deprived workers in order to enrich the bourgeoisie, Communist union leaders within the state sector recognized their position of power and were not about to go into opposition at the grass roots level, where the coalition therefore continued.

The technocratic-Communist coalition inside the modernization network began to close ranks in the face of opposition. Following a set of state budget cuts on modernization projects in 1948, the Communist head of the CGT electrical workers' union joined the head of France's largest electrical equipment and public works combine on a publicity tour of "the land of the kilowatt." In the publicity brochure produced after the tour, the collective authors vaunted the camaraderie of the construction workers, who labored up to ninety-four hours per week as the pioneers of the emerging new France. At the Bort dam site, "In less than five years, men correct[ed] the work of millenia." Peasant resistance to the dams was written off simply as a result of "medieval fear and superstition." The pre-Cold War vision of the working class was underlined: "Social barriers are abolished, unknown. The hierarchy exists, [it is] normal and natural, but as a function of value, competence, and love for the task ahead."[70] Similarly, when the peasants at Tignes in Savoy desperately fought the construction of a hydroelectric project which would annihilate their village, Communist unionists and modernist industrialists combined to suppress opposition (using no small number of riot police who occupied the village for several years) and forge ahead with the project.[71] Of the incident and the social costs of modernization, one journalist wrote,

The basic and significant fact was not the sacrifice of a village for the production of energy, but that it was not even discussed. Between man and production, the debate is resolved in advance, and always in the same way. This automatic reaction and the inhibitions it places on our spirits shows clearly that productivism is not for us a rational issue but a collective myth. [Indeed,] religious dogma has been replaced by economic dogma.[72]

The imagery of technological modernization was indeed mythological. Enamored by sheer scale and by an emerging cult of mathematical economic analysis, technocrats began to demonstrate the economic necessity for ever-larger facilities, ignoring issues of system reliabilities inherent in smaller scale. The new dams emerged as icons of the new France, though Tignes today represents only two percent of France's generating capacity. From the minimalist elegance of Tignes's dam to the Romanesque arches at Roselend and the new series of coal plants, massive transmission lines spider-webbed a new France together. The modernization coalition was succeeding, and each completed project became another signpost on the path to a technologized future.

At the same time, mathematical economists within the modernization networks developed new mathematical models to justify cheap power and rail rates to the largest consumers. Increasingly, technological artifacts and economic equations became reified from their socio-political context; facilities became material proof of the success of the modernization program, and workers—even white-collar technicians—became mere factors of production. This decontextualization of people, artifacts, and equations was not unlike the decontextualization of scientific facts as they become part of accepted canons.[73] Planning models began to take a mythological free market as a reference standard for all projects, regardless of industrial or market structure, or the security of material inputs. For example, guided by formulas which projected the price of oil falling well into the 1980s, the power company had several thousand megawatts of oil-fired power plants under construction when the crisis hit in 1973. Worse, experts had calculated that the total capital costs for fuel-conserving technology could not be justified when oil prices were low, so France was saddled with a set of inefficient plants as prices trebled. Mathematics became a language not only of description and analysis, but also of obfuscation and inaccessible power, performing the useful task of insulating the technocracy from outside criticism.[74]

Communists shared many of the neo-positivist predilections of the technocrats, part of an ideological tradition within orthodox Marxism.[75] Indeed, during the years of political isolation, French Communists began to adopt the old interwar imagery of the link between mass production and popular consumerism. For Communists within the modernization networks, this old rhetoric represented a plausible future; for Communists outside, it represented a mode of criticism against a stingy capitalist order. Those inside continued a heroic proof by production, using output statistics as a mode of defense against parsimonious régimes. This meant that notions of "democratic management," once vaunted by the CGT, were removed from the agenda in favor of a deference to technocratic authority, now judged to be the surest guarantee of productive success. An implicit prospective narrative thus began to emerge for the Communists: deferred consumption in the present would foster larger industrial structures and later popular consumerism. The director of the power company characterized the

Communist attitude on the shop floor as one of, "Work hard, follow orders, and collect your growing paycheck."[76] Making the consumerist agenda more explicit—characterizing the use of the larger paycheck—Marcel Paul said, "The washing machine is now an instrument of class struggle."[77] The Communists never discarded the dogma that the workplace was to be the flashpoint of class struggle, yet by supporting a depoliticized consumerist mentality, they were undercutting their own position.

During the 1950s, therefore, many businessmen, technocrats, and trade unionists developed shared visions of a technologized future, offering parallel scenarios about how to get there. For these constituencies, a technologically-based growth economy became a central political concern, yet the vision of growth was itself largely depoliticized. In the mid-1950s, the productive successes were obvious and opponents were rendered increasingly mute. By the late 1950s, the need to complete infrastructural development and to wrest control of the state from the provincial traditionalists who dominated Parliament remained the only barriers to realizing the modernists' vision of the future. Both of these tasks were completed by 1960, and the emerging Gaullist state—one which gave both big business and Communist labor an institutional presence—symbolized their fulfillment.

De Gaulle himself strongly believed in technological imagery, particularly for nationalist uses. It can be argued that he opted to pull out of Algeria because a dirty little war precluded France's emergence as a great power. The development of nuclear weapons, the archetypical icon of the marriage between technology and nationalism, would for de Gaulle and others compensate for the disaster of lost empire. One may muse over the utility of nuclear weapons on a small continent, particularly when deployed against a power with whom de Gaulle was seeking friendly relations, but image was more significant than the function. Indeed, de Gaulle initiated a wave of technological projects which made far more sense as objects of display, because their economics were dubious; the Concorde and the French design of nuclear reactors were only the most obvious artifacts of this sort.[78]

After de Gaulle's departure, his successors extended the logic of technology-as-display.[79] The high-speed Paris-Lyon train is indeed fast, but it is rather expensive for popular transport and relegates many towns without appropriate stations to oblivion. The design of the Pompidou museum uses the architectural æsthetics of an oil refinery to minimize the art it displays. The "Cité of Sciences and Industry" at La Villette decontextualizes science and technology with its infantalizing, Mr. Wizard approach to visitor participation. The world's most massive atomic power system, replete with a complete nuclear fuel cycle, is vastly oversized for the task, but it remains the object of adoration for depressed utility engineers worldwide. At times, technological display is used not to ascribe new meanings, but to annihilate them: an earthquake room at La Villette placed earth tremors on a program and thus destroyed the disconcerting sense of an earthquake's total unpredictability. Similarly, a proposal for a massive mirror placed in orbit around the earth to commemorate the French revolutionary bicentennial ascribed no meaning to the revolution itself. (Opposition from astronomers outside of France stopped the project.) There was (and is) rarely political opposition to these projects; indeed, the Communists' central criticism of the nuclear program was that it was too small.

De Gaulle's rise to power similarly reinvigorated the planning apparatus. He appointed Pierre Massé, one of France's top technocrats, as the Commissioner of the Plan. Massé had once claimed that with his models, he had successfully rendered quality "quantitative and calculable."[80] He began his work on the Fourth Plan (to run from 1962 to 1965) in 1959, on the assumption that France had successfully completed her infrastructure-building, so that French state policy could now begin to respond to consumers' needs by consciously redirecting a portion of national income flows from investment and toward households. The political benefit of this to the régime was incalculable and clear; the austerity economics of the Fourth Republic were history.

In the 1960s, state-sector technocrats encouraged the development of consumer-goods industries and urged its nationalized entities to do so as well. In this fashion and with the encouragement of the CGT, Électricité de France (EDF) not only helped to promote the infant appliance firm Moulinex, it also began to promote wider domestic electrical use, offering financial help and technical oversight for expanding residential electrical wiring.[81] Emphasis in transportation policy shifted from railroads to cars. Coming from firms which had legitimated themselves through the 1950s by reference to heroic-scale technological artifacts, state-fostered consumerism reflected its roots in reified technology, from the nuclear strike force of the publicly-owned arms sector to the "all electric home" promoted by EDF. Technology had liberated France from want. Now it was time to liberate the nation from the danger of foreign domination and French women from the travails of housework. Seeking self-justification and political aggrandizement, modernists began to divorce technology from its thickly political and social provenance and attribute entirely new and reified meanings to artifacts.[82] The invention of a new context for artifacts was a critical part of the new marketing mentality, and the terrain of redefined signs implied a new cultural frame, characterized by mass media, rapid transit, rationalized work, and the kitchen-as-clinic—a push-button world commanded by a technologized populace, the alienated world of Godard's *Alphaville*.

Thus, the imagery of EDF and appliance company advertising in the 1960s echoed that of the 1920s, yet the imagery finally had a credible material base. The revised scenarios built upon technological infrastructures and the promise of upward social mobility. Discount retailers touted a plethora of new appliances and domestic technologies, all designed to make the housework easier for the *maîtresse de maison*. She was to become a genuine sartorial and culinary *technicienne*. For the first time since before the First World War, French women were entering the labor force in large numbers, and advertisements proclaimed that the high-tech home would make women's "natural" task, housework, easier.[83] As gift-giving occasions such as Christmas and Mothers' Day became more resplendent, many of the promotional efforts were aimed less at women than at husbands, who could participate in appliance-giving rituals as a means to underscore and reinforce the intrafamilial division of labor and gender roles. The modern hubby gave his wife a new, clean-lined mangle as much to demonstrate his role as provider and hers as housewife as actually to iron clothes. Most importantly, however, the subtext of the advertising and of the mass consumption nexus generally annilhilated politics in the public and private spheres alike.

While the revolt of May 1968 placed the new consumerist ethos under intense scrutiny and criticized the sterility of the new social order, it also signalled that the apolitical technological society had finally emerged. As they demanded a better quality of life in the place of more goods, the *soixante-huitards* knew all too well that the cults of technology and quantity of goods had reduced political language to the reproduction of empty mythologies. Material success based in technological mastery seemed to render difficult the old debate over the division of the economic pie. The new technocratic élite had shifted public discourse about political-economic choices to narrow debates over what centralized, expert-run, and monopoly-owned technological systems would look like.[84] After the failure of the revolt, the discourse of the political class in France degraded to disagreements over tactics to achieve the unquestioned goal of economic growth driven by reified technology, managed by technocrats. Firms were to be large, centralized, expert-managed, and rigidly hierarchical internally. In political terms, debate centered only upon the modes of financing such firms—equity-financed if privately owned and debt-financed if publicly owned. Aside from a few Catholics and *gauchistes*, the scenarios, or prospective narratives of left and right, modernist and traditionalist, had fully converged: France gained the high-tech world of prosperity, and

MECHANICAL DREAMS IN TWENTIETH-CENTURY FRANCE 71

public debate over collective choices lost. There is little wonder why pundits now write of a French polity that is calmly divided between two essentially centrist parties. French society constructed a common technological sign system, and in the process emaciated the language needed to discuss the values behind it. The *citoyens* and *citoyennes* of 1989 faced a new Bastille at Creys-Malville.

State University of New York at Albany

NOTES

1. For an incisive review of the Eiffel Tower as a technological and political artifact, see Miriam R. Levin, "The Eiffel Tower Revisited," *French Review* 62:6 (May 1988): 1052-1064.
2. "Do Machines Make History?," *Technology and Culture* 8:3 (1967):336, cited in John M. Staudenmaier, S.J., *Technology's Storytellers: Reweaving the Human Fabric* (Cambridge, Mass.: MIT Press, 1985), p. 164. See also David F. Noble, *Forces of Production: A Social History of Industrial Automation* (New York: Knopf, 1984), pp. 144-147. In the latter work, this view is characterized as a form of naive neo-Darwinism.
3. For an astute summary of the historiography of technology, see T. Misa, "How Machines Make History, and How Historians (and Others) Help Them to Do So," *Science, Technology, and Human Values* 13:3 & 4 (Summer-Autumn 1988): 308-331.
4. Arnold Pacey, *The Culture of Technology* (Cambridge, Mass.: MIT, 1985), and Jeffrey Meikle, *Twentieth Century Limited: Industrial Design in America, 1925-1939* (Philadelphia: Temple University Press, 1979).
5. David S. Landes, *The Unbound Prometheus: Technological Change, 1750 to the Present* (Cambridge: Cambridge University Press, 1967), and Alfred F. Chandler, *The Visible Hand: The Managerial Revolution in American Business* (Cambridge, Mass.: Harvard University Press, 1977).
6. Nathan Rosenberg takes a similar tack, discussing unilinearity most explicitly in his discussion of "bottlenecks," see "The Direction of Technological Change," *Economic Development and Cultural Change* 17 (October 1969): 1-14.
7. Thomas P. Hughes, "The Evolution of Large Technological Systems," in Wiebe Bijker, Thomas P. Hughes, and Trevor Pinch, eds. *The Social Construction of Technological Systems* (Cambridge, Mass.: MIT Press, 1987), pp. 51-82; see also his *Networks of Power: Electrification in Western Society* (Baltimore, Md.: Johns Hopkins University Press, 1983); John Law, "On the Social Explanation of Technical Change: The Case of Portuguese Maritime Expansion," *Technology and Culture* 28:2 (1987): 227-252, and "Technology and Heterogeneous Engineering: The Case of Portuguese Expansion," in Bijker, Hughes, and Pinch, *Social Construction*, pp. 111-134.
8. Perhaps the best example for Hughes in terms of internal technical logics is his recognition of the non-storable character of electricity, ineluctably demanding that utilities produce the exact quantity of power needed at any given moment. Production capacity in power utilities thus has to be equal to maximum rather than long-term average demand levels, in turn demanding a highly capital-intensive industry. See *Networks*, chapter 1; also, UNIPEDE, *Méthodes de prévision de consommation à moyen terme* (Paris: UNIPEDE, 1972), p. 9. At the same time, one is still left with the sense that by their inexorability, technical constraints are ontologically prior to lower-order political

or social ones, parallel to the ranking of knowledge by the French Encyclopædists in the eighteenth century; see Robert Darnton, *The Great Cat Massacre and Other Episodes in French Cultural History* (New York: Basic Books, 1984), chapter 5: "Philosophers Trim the Tree of Knowledge: The Epistemological Strategy of the *Encyclopédie.*"

9. Hughes, "Evolution," pp. 71 ff.

10. John Law and Michael Callon, "The Life and Death of an Aircraft: A Network Analysis of Technical Change," in Weibe Bijker and John Law, *Constructing Networks and Systems: Case-Studies and Concepts in the New Technology Studies* (Cambridge, Mass.: MIT Press, forthcoming).

11. A possible solution to the problem of anti-system systems might be to develop Michel Foucault's anti-causal "genealogies" into socio-technical constructs; see Patricia O'Brien, "Michel Foucault's History of Culture," in Lynn Hunt, ed., *The New Cultural History* (Berkeley: University of California Press, 1989), pp. 33-35.

12. Perhaps this is rooted in the fact that *Networks* deals only with developing systems (up to 1930), not with mature ones. One is left wondering, nonetheless, if the best choices were as obvious and singular as they appear in retrospect.

13. Hughes, *Networks,* introduction.

14. *Culture technique* 3 (special number, September 1980); Ruth Schwartz Cowan, *More Work for Mother* (New York: Basic Books, 1983); and Noble, *Forces.*

15. Langdon Winner, "Do Artifacts Have Politics?," *Daedalus* 109 (1980): 121-136, and *Autonomous Technology: Technics-Out-of-Control as a Theme in Political Thought* (Cambridge, Mass.: MIT Press, 1977).

16. Shoshana Zuboff, *In the Age of the Smart Machine* (New York: Basic Books, 1988), chapter 4, and Margaret Hedstrom, "Automating the Office: Technology and Skill in Women's Clerical Work, 1940-1970," Ph.D. dissertation, University of Wisconsin-Madison, 1988, chapter 4, have argued that new technologies often destabilize existing social relations and can escape the control of their authors. One is reminded of literary critics' arguments about how texts can become autonomous from their authors.

17. Hayden White, *Tropics of Discourse: Essays in Cultural Criticism* (Baltimore, Md.: Johns Hopkins University Press, 1978), pp. 236-238.

18. The possible links between this approach and those of literary criticism and symbolic anthropology are obvious and will be discussed below. The most recent version of Bijker's evolving argument is, "The Social Construction of Fluorescent Lighting, Or How an Artefact was Invented in Its Diffusion Stage," in Bijker and Law, *Constructing Networks and Systems.*

19. The simplest way to use this approach is to deconsruct the functions of an artifact and, once the interests have been "read" from it, find the artificer and his/her interests.

20. Bijker and many other sociologists rely on what Joan W. Scott has termed "objectively determined interests." She and the present author would instead seek to discern interests as being discursively produced through social, material, rhetorical, and symbolic interactions; see Scott, *Gender and the Politics of History* (New York: Columbia University Press, 1988), p. 5.

21. This cessation of contention could be represented within a Marxist framework as the successful reproduction of existing social relations; this point is explicitly argued by Bryan Pfaffenberger, below.

22. Bruno Latour, *Science in Action* (Cambridge, Mass.: Harvard University Press, 1977), and Madeleine Akrich, "How Can Technical Objects be Described," conference paper, Twente colloquium, 3-5 September 1987.

23. For a basic summary of resource mobilization theory, see J. Craig Jenkins, "Resource Mobilization Theory and the Study of Social Movements," *Annual Review of Sociology* 9 (1983): 527-553. For a masterful application of resource mobilization theory, see Doug McAdam, *Political Process and the Development of Black Insurgency,*

1930-1970 (Chicago: University of Chicago Press, 1982).

24. Byran Pfaffenberg, "The Social Construction of the Personal Computer, Or, Why the Personal Computer Revolution Was No Revolution," *Anthropological Quarterly* 61 (1988): 39-47; see also Pfaffenberger, "Fetishized Objects and Humanized Nature: Towards an Anthropology of Technology," *Man* 23 (1988): 236-252.

25. This problem parallels that of Lynn Hunt in *Politics, Culture, and Class in the French Revolution* (Berkeley: University of California Press, 1984), chapter 3. Without, say, a generation of meaning by the internal structures of a sign system (in classically structuralist fashion), the parameters for decoding signs-as-perceived implicitly rest on the inexplicit, shared presumptions of analyst and reader.

26. It is at this juncture that a semiotically expanded version of Law's "heterogeneous engineering" or of Latour's translations might be able to encompass concurrent adjustments of meanings and of real artifacts, see Law, "Technology and Heterogeneous Engineering," and Latour, *Science in Action*.

27. Pfaffenberger, "Personal Computer," p. 22.

28. Roland Barthes, *Mythologies* (New York: Hill and Wang, 1972), pp. 114 ff.

29. *Ibid.*, p. 121.

30. Pierre Lemonnier goes a step further, arguing that one cannot conceptually divorce form or style or function—indeed, that form has a functional role: "Bark Capes, Arrowheads, and Concorde: On Social Representations of Technology," in I. Hodder, ed., *The Meaning of Things: Material Culture and Symbolic Expression* (London: Unwin Hyman, 1989), pp. 156-169.

31. The work of Hayden White and Theodore Sarbin (both at the University of California, Santa Cruz), in the history of consciousness and psychology, respectively, is promising in this respect. See particularly Theodore R. Sarbin, ed., *Narrative Psychology: The Storied Nature of Human Conduct* (New York: Praeger, 1985). For a review of scholarship in constructivist psychology, see J. M. Sperry, "Structure and Significance of the Consciousness Revolution," *Journal of Mind and Behavior* 8 (Winter 1987): 37-61. Robert Neimeyer writes that personal construct psychology "depart[s] from the environmental determinism of radical behaviorist formulations and the intrapsychic determinism of classic psychoanalysis" in *The Development of Personal Construct Psychology* (Lincoln: University of Nebraska Press, 1985), p. 2.

32. For a capsule summary of the method, see James C. Mancuso and Theodore R. Sarbin, "The Self-Narrative in the Enactment of Roles," in T. R. Sarbin and Karl Scheibe, eds., *Studies in Social Identity* (New York: Praeger, 1983), pp. 234-253, especially 235-237.

33. Most recently, see Richard Harvey Brown, "Positivism, Relativism, and Narrative in the Logic of the Historical Sciences," *American Historical Review* 92:4 (October 1987): 908-920. Oral historians have used this approach for some time; see, for example; Kim Lacy Rogers, "Memory, Struggle, and Power: On Interviewing Political Activists," *Oral History Review* 15 (Spring 1987): 165-184. See also Hunt, *Politics, Culture, and Class*, p. 88.

34. While one may rightfully criticize, as Clifford Geertz does, the mainstream Parsonian stress upon stasis and the maintenance of social order as a critical tool, this does not preclude the possibility that historical actors might have sought stasis or the annihilation of change; see Clifford Geertz, *The Interpretation of Cultures* (New York: Basic Books, 1973), pp. 142-144.

35. On this subject in the psychology literature, see Jean Matter Mandler, *Stories, Scripts, and Scenes: Aspects of Schema Theory* (Hillsdale, N.J.: Erlbaum, 1984), chapters 2 and 3.

36. Madeleine Akrich, "The De-scription of Technical Objects," in Bijker and Law, *Constructing Networks*.

37. A successful invention of tradition can supplant this rule. Vestigial wood grain

patterns still adorn station wagons and suggest the "tradition" of the 1930s "woodies"; and baby shampoo has remained unchanged for decades, thereby encouraging intergenerational product association. Neither artifact need be reinvented or redeveloped. On the invention of tradition in another context, see Eric Hobsbawm, "Mass Producing Traditions: Europe, 1870-1914," in Eric Hobsbawm and Terence Ranger, eds., *The Invention of Tradition* (New York: Cambridge University Press, 1983), pp. 263-307.

38. See Pfaffenberger, "Personal Computer," pp. 16-18.

39. Such was the language of André Gorz, *Ecology and Politics* (Boston: South End Press, 1980), chapter 3.

40. See, for example, the essays in Nathan Rosenberg, *Inside the Black Box: Technology and Economics* (New York: Cambridge University Press, 1982).

41. Notable exceptions are Cowan, *More Work for Mother*; Noble, *Forces*; and Zuboff, *Smart Machine*.

42. Sherry Turkle, *The Second Self: Computers and the Human Spirit* (New York: Simon and Schuster, 1984), p. 13.

43. Hunt, *Politics, Culture, and Class*, p. 13.

44. See, for example, Michael L. Smith, "Selling the Moon: The U.S. Manned Space Program and the Triumph of Commodity Scientism," in T. J. Jackson Lears, et al., eds., *The Culture of Consumption* (New York: Pantheon, 1983).

45. Jean Baudrillard, *Le Système des objets: La Consommation des signes* (Paris: Gallimard, 1968), pp. 21-23.

46. Martin Fine, "Hyacinthe Dubreuil: Le Témoignage d'un ouvrier sur le syndicalisme, les relations industrielles et l'évolution technologique de 1921 à 1940," *Le Mouvement Social* 105 (1977), entire issue; and Martine Martin, "La Rationalisation du travail ménager en France dans l'entre-deux guerres," *Culture technique* 3 (1980): 161; Martin, "Ménagère: Une Profession? Les Dilemmes de l'entre-deux-guerres," *Le Mouvement Social* 140 (July-September 1987): 89-106; and Martin, "Femmes et société: Le Travail ménager (1919-1939)," thesis, University of Paris VII, 1984, chapter 3.

47. Michelle Perrot, "Histoire de la condition féminine et histoire de l'électricité," in *L'Électricité dans l'histoire, problèmes et méthodes: Actes du colloque de l'Association pour l'histoire de l'électricité en France, Paris, 11-13 octobre 1983* (Paris: Presses Universitaires de France, 1985), pp. 175-184; and Françoise Werner, "Du Ménage à l'art ménager: L'Évolution du travail ménager et son écho dans la presse féminine française de 1919 à 1939," *Le Mouvement Social* 116 (1984). Also, Charles S. Maier, "Between Taylorism and Technocracy: European Ideologies and the Vision of Industrial Productivity in the 1920s," *Journal of Contemporary History* 5 (1970).

48. Robert L. Frost, *Alternating Currents: Nationalized Power in France, 1946-1970* (Ithaca, N.Y.: Cornell University Press, forthcoming), chapter 1; and Patrick Fridenson, *Histoire des usines Renault* (Paris: Éditions du Seuil, 1972), chapter 5. On the study circles, see Richard F. Kuisel, *Capitalism and the State in Modern France* (New York: Cambridge University Press, 1981), chapter 3; and Philippe Bauchard, *Les Technocrates au pouvoir* (Paris: Arthaud, 1966), chapter 2.

49. Yves Cohen, in "Management, Organization, and Production in French and American Automobile Industries Between the Two World Wars: Some Aspects," a paper delivered to the annual meetings of the Society for the History of Technology, Cleveland, October, 1990, cites this wariness even on the part of the boldest innovators in French industry, the auto makers. Aimée Moutet consistently argues that most French industrial sectors were too small to achieve the necessary scale for mass production; see in particular, "Une Rationalisation de travail dans l'industrie française des années trente," *Annales E.S.C.* 42:5 (September-October 1987): 1087. The important point here is not whether or not in fact mass markets were lacking, but that entrepreneurs perceived them to be.

50. Patrick Fridenson, "Les Expériences de la rationalisation dans la France des années 1920," *Sciences Sociales et Santé* 6:3-4 (November 1988): 51-54; Aimée Moutet, "L'Introduction de travail à la chaîne," *Histoire, Économie et Société* (January-March 1983); and Ernest Mercier, "Le Redressement Français," speech to the Comité National d'Études, 28 January 1926, published as a pamphlet (Paris: Comité National d'Études, 1926).

51. French manufacturers replicated the problems that David Hounshell discusses in *From the American System to Mass Production* (Baltimore, Md.: Johns Hopkins University Press, 1987), chapter 2, where he characterizes the European style for the Singer works in the late nineteenth century: many fitters and no standardized parts.

52. Loose document and attached form letter from Paul Breton, head of Salon des Arts Ménagers, 15 May 1928, Archives Nationales code 850023/3.

53. This was particularly the case for refrigerators: Frigéco (General Electric designs manufactured to specifications by Thomson, SA, in France), Frigidaire, and Norge all essentially imported American ad copy for their French ads. Norge's advertisement of 1935 for the French market even pictured American products.

54. For example, Birnam-Lutra offered a floor polisher in 1928 which in its advertising text bragged of its industrial strength and ability to do the job (an appeal to the female users' practicality), yet presented a line drawing of a well-coiffed woman in spike heels and short skirt deftly manipulating a 50+ kg. machine with one hand (an appeal to the husband, who might fear that his wife would be de-feminized by industrial machinery): advertising copy in *Catalogue du Salon des Arts Ménagers*, 1928 An côte 850023/6, n.p.

55. Anonymous, "Budgets ouvriers: Le Travail et la famille," *Dossiers de l'Action Populaire* (25 October 1923), p. 3. Jean Fourastié et Françoise Fourastié, "Le Gendre de vie," chapter 13 in Alfred Sauvy, *Histoire économique de la France entre les deux guerres*, vol. 3 (Paris: Economica, 1984), pp. 214-222. In 1928 a regular stand-up vacuum cleaner by Mors cost 855 francs and a skilled male textile worker made 5 to 6 francs per hour: AN série F22 191 (for wage levels, which for unskilled workers outside Paris hovered at about 2 to 3 francs per hour); vacuum cleaner price from Mors prices listed in 1928 Salon des Arts Ménagers catalog, in Archives Nationales series 850023/65.

56. Interviews by Gaston Picard, "Mesdames, êtes-vous 'Arts-Ménagères?'," *L'Art Ménager* (March 1927), p. 24.

57. Peggy Phillips, "The Lonely House: Paris and France," *Proceedings of the Annual Meeting of the Western Society for French History* (1986), and communication with author.

58. Some of the Popular Frontist language was strictly *anti*-technological, blaming industrial rationalization for the high rates of unemployment. Indeed, the mix of social-technocratic and anti-technological rhetoric within the Popular Front is perhaps symbolic of the divisions within the coalition.

59. For the best discussion of this productivity boom in the airframe construction industry, see Herrick Chapman, *State Capitalism and Working Class Radicalism in Twentieth Century France: Industrial Politics in the French Aircraft Industry, 1928-1950* (Berkeley: University of California Press, 1990), pp. 164-165.

60. Comité Général d'Études (C.G.E.), "Pour une nouvelle revolution française," *Les Cahiers Politiques* 5 (January 1945) 11.

61. Philippe Robrieux, *Histoire intérieure du Parti communiste* (Paris: Fayard, 1981), vol. 2, pp. 81-85; and Charles Tillon, *On chantait rouge* (Paris: Laffont, 1977), p. 445.

62. On the CGT's analysis of managers, see Annie Lacroix, "La CGT et action ouvrière de la liberation à mai 1945," *Revue d'Histoire de la Deuxième Guerre Mondiale* 116 (October 1979); and Darryl O. Holter, "Miners against the State," doctoral dissertation, University of Wisconsin- Madison, 1980, pp. 231-242.

63. Part of this fixation with large-scale, centralized systems is also probably rooted in the training of French engineers, whose Cartesian mathematical approach makes them far better at large "drawing board" conceptions and weak at more mundane "detail" work.

64. The program documents for the Monnet Plan, authored in 1946, bear this out. In the electricity program, for example, dam projects were set which used vast quantities of capital, construction machinery, concrete, and labor, none of which were in sufficient supply. See France-Commission de la Modernisation de l'Électricité, *Rapport* (Paris: Imprimerie Nationale, 1946); Henry Rousso, ed., *De Monnet à Massé* (Paris: Éditions de la CNRS, 1986); and Alan S. Milward, *The Reconstruction of Western Europe, 1945-51* (Berkeley: University of California Press, 1984), chapters 1 and 2.

65. Guy Bouthillier, "La Nationalisation du gaz et de l'électricité en france," thesis, Univerisity of Paris, 1968; Frost, *Alternating Currents*, chapter 2. *Communisant* authors argue otherwise, claiming that business put up a genuine fight; see Annie Lacroix, "La Nationalisation de l'électricité et du gaz," *Cahiers d'Histoire de l'Institut Maurice Thorez* 6 (new series 34; January 1974). Symbolic resistance does not mean, however, that opponents conceded permanent defeat; René Courtin wrote: "It is the bourgeoisie who have betrayed the country. We will therefore nationalize. It is a political necessity. In fifteen years, when a new bourgeoisie emerges which is worthy, we will denationalize"; cited in interview with André Philip, in Sven Nordengren, *Economic and Social Targets in Postwar France* (Lund, Sweden: Belingska Boktryckriet, 1972), p. 240.

66. The concept of "technological display" has been fruitfully developed in Michael L. Smith, "Selling the Moon."

67. Kuisel, *Capitalism and the State*, chapter 5.

68. Jules Moch, a Polytechnique Socialist leader, was most explicit about this; see his series of pamphlets and articles, from *Socialisme et rationnalisme* (Paris: Librarie du Parti, 1924), through "Reflexions sur les socialisations," *Les Cahiers politiques* (clandestine), 8 (March 1945) [pseud. J.M.], to *Guerre aux Trusts* (Paris: Editions de la Liberté, 1945).

69. For a full discussion of the Communists' position, see Frost, *Alternating Currents*, chapters 2 and 3.

70. Le Comité pour l'Équipement Électrique Français, "Cinq jours au pays du kilowatt," pamphlet (Paris: CPEEF, 1949), pp. 9 and 40.

71. For a complete rendition of the Tignes incident, see Robert L. Frost, "The Flood of 'Progress': Technocrats and Peasants at Tignes (Savoie), 1946-1952," *French Historical Studies* 14:1 (Spring 1985): 117-140.

72. G. Charbonneau, "Maintenant que Tignes est oubliée *Réforme* a choisi d'en parler," *Réforme* (28 June 1952), p. 6.

73. Bruno Latour and Steve Woolgar, *Laboratory Life: The Social Construction of Scientific Facts* (Beverly Hills, Calif.: Sage, 1977), pp. 105-107.

74. Members of the Senate (Conseil de la République) became particularly incensed at the obfuscatory nature of the technocrats' equations during debate over the enactment of a marginal cost high tension electrical power tariff; see Conseil de la République, Commission de la Production Industrielle, "Rapport No. 418," pamphlet, annexe au procès verbaux du Conseil, séance du 21 juillet 1954 (Paris: Imprimerie Nationale, 1954), pp. 2-4.

75. In this context see D. MacKenzie, "Marx and the Machine," *Technology and Culture* 25 (July 1984): 473-502, and Winner, *Autonomous Technology*, pp. 65-75. Louis Puiseux, a former Électricité de France official, stated a similar perception based on his own experience in an interview with the author, Paris, 18 February 1981.

76. Interview by author with Roger Gaspard, Paris, 23 June 1981.

77. Paul's report to the 25th CGT National Congress, June 1955, cited in

MECHANICAL DREAMS IN TWENTIETH-CENTURY FRANCE 77

Jean-François Picard, Alain Beltran, and Martine Bungener, *Histoire(s) de l'EDF* (Paris: Dunod, 1985), p. 375.

78. On economics of the Concorde (albeit in Britain), see Peter Geoffrey Hall, *Great Planning Disasters* (London: Weidenfeld and Nicolson, 1980), chapter 4; on the French nuclear design, see Picard, *Histoire(s)*, pp. 185-203; and Frost, *Alternating Currents*, chapter, 5.

79. The term is from Michael L. Smith, "Selling the Moon."

80. Électricité de France, *Dix ans de progrès* (Paris: EDF, 1957), p. 19.

81. Letter from Marcel Paul (head of CGT-Fédération Nationale de l'Énergie (FNE)) to Roger Gaspard (CEO of Électricité de France), 2 February 1960, CGT-FNE archives, Pantin, dossier: "Entreprises privées"; see also *Le Moniteur professionnel de l'électricité* 73 (July 1961); H. Pottier, "L'Industrie du matériel électro-ménager," *Revue Française de l'Énergie* 134 (November-December 1961): 162-170; Editorial, "Un tournant à l'Électricité de France," *Revue Française de l'Énergie* 137 (March 1962): 299; and Marcel Boiteux, "Principes de la politique commerciale de distribution d'Électricité de France," *Bulletin d'Information des Cadres d'Électricité de France,* article no. 30 (December 1959), pp. 1-6.

82. Ironically, even New Left writers fed into this emerging ideology: writers such as André Gorz, in *Strategy for Labor* (Boston: Beacon, 1964), and Serge Mallet, *The New Working Class* (Nottingham, England: Bertrand Russell Peace Foundation; Spokesman Books, 1975), developed "new working class" theory, which argued that the technology-driven automated workplaces of the post-industrial age had liberated workers from drudgery and compelled them to raise new demands about the quality of worklife and of consumer products.

83. Drawn from advertisements on file at EDF offices, Centre de Documentation Murat-Messine, Paris.

84. One can argue that the centralized, expert-run synthesis of state and economy has its origins with Colbert, running through Necker, Robespierre, and Étienne Clémentel, finally to Pierre Massé and Pierre Mesmer. Nonetheless, this interpretation would exclude a large part of the history of the Third Republic, in particular, the constant victories of provincial politicians against Parisian experts and technocrats (a common truism about the Third Republic is that it represented the victory of the provinces over Paris), as well as the major strains of French industrial historiography.

PART II

SOME PROPOSED SOLUTIONS

PAUL T. DURBIN

MARXISM AND THE DEMOCRATIC CONTROL OF TECHNOLOGY

The question I raise here is a simple one. Assuming anything of a democratic nature can be done to control the future direction of technological development, which of two prominent approaches—progressive liberalism or Marxist revolution—offers more hope?

At least twice before[1] I have attempted to defend a progressive liberal social philosophy of technology against Marxism—only to have my dismissal of that view dismissed as glib[2] or utopian.[3] Albert Borgmann,[4] in spite of being much more careful in analyzing the version of Marxism he chose to attack, has met much the same fate; what he has been accused of, in rebuttal, is a failure to deal with the latest version of Marxism, the one claimed to be the most authentic or the most appropriate for our technological era.[5]

I am not here primarily concerned with the latest iteration of Marxist theory.[6] Indeed, the collapse of Communism as both an ideology and a system of rule in Eastern Europe and the former U.S.S.R. makes these revisions seem irrelevant. What I am concerned with is the real-world problem Marxists have long claimed to address—the social problems spawned by capitalism. Like Karl Marx[7] himself, I am not so much interested in philosophical disputes over the best interpretation of the world; I am interested in changing our world for the better. I am, with John Dewey[8] and George Herbert Mead,[9] a social meliorist; and I am convinced that a progressive, liberal-democratic politics is more likely to deal effectively with the threats of technology than the politics of international social revolution.

THE REVOLUTIONARY SOLUTION TO SOCIAL PROBLEMS ASSOCIATED WITH TECHNOLOGY

I have proposed elsewhere[10] a list of ten types of social problems that beset contemporary high-technology society. The problems range from the nuclear arms race to commercialization of traditional high culture, from ecological catastrophes and genetic engineering to boredom in high-technology jobs and alienation in family life in today's sprawling urban centers. But at the center of my list is growing technoeconomic injustices, and especially the increasing disparity between the haves and the have-nots—whether these are intranational, between socioeconomic classes in high-technology economies, or international, between developed and so-called developing nations.

It is this problem that Marx, and Marxists ever since, have emphasized. In my view, any interpretation of Marx that does not focus primarily on the class struggle between, on one hand, those who control the means of production appropriate to a given stage in the dialectic of history, and, on the other hand, the exploited workers who actually produce economic wealth is not within the mainstream of Marxist theory. Furthermore, any authentic Marxist ought to say that none of the other problems of technological society that I list will be solved until the class struggle is resolved worldwide.

Why is this? There would seem to be an obvious link between the economic issue—especially if interpreted in class-struggle terms—and all the other issues: the nuclear economy obviously; industrial and consumption-driven wastes; the temptation of the haves to use high-tech surveillance methods, and perhaps eventually genetic intervention, to keep the exploited have-nots in line or to mold them for particular sorts of work; bribes for workers to induce them to accept hazardous or mind-numbing jobs; worker alienation carrying over into family life, or even leading to its breakdown; schools turned into corporate training grounds without attention to their traditional role of educating responsible citizens; politics turned into media manipulation, frustrating true democracy; the arts no longer critical of society but corporation-dominated.

Several common interpretations of what is going on here need to be dispatched quickly. When today's students come in contact with Marxist views on the impact of economic power on social problems, they often think of it as the exercise of raw economic manipulation. Wealthy individuals, high-level corporate managers, politicians in league with the wealthy and managerial classes, can simply do as they will. If it means profit for them, they can start wars or keep cold wars going indefinitely. Similarly, critics often take Marxists to be saying that leaders of the capitalist exploiting class act in conscious concert to control education or the media. In turn, cynics interpret capitalist exploiters as pure and simple greedy men who will do anything, no matter the effects on workers or on the environment, if it means more short-term profits for themselves. And, of course, short-term profits turn into long-term capital investments, and the cycle goes on.

None of these interpretations is necessarily or entirely false. No doubt leading capitalists do exercise raw economic power, do sometimes act in collusion in ways that seem to amount to conspiracy (or monopolistic practices), and can be as greedy as anyone else in society. But none of this is the point of the Marxist claim that class divisions pitting capitalists against workers are the root of all social ills in our technological society—or in any previous version of capitalist society. According to Marxist theory, it is not the individual motives of capitalists, singly or acting in concert, that explain why class-division disparities between capitalists and workers lead inevitably to toxic wastes, hazardous workplaces, and boring high-technology jobs. In this view, what makes these social problems insoluble until exploitation ends is that capitalism is a wholesale ideological superstructure erected on the base or substructure of capitalist-era modes of production. Our entire way of life, all our social relations, not only at work but in the home and everywhere else, are intelligible only in terms of the ideology of capitalism (or, in the present view, techno-capitalism).

Eugene Genovese provides a telling picture of how this works in his interpretation of life and its accompanying worldview in the slaveholding society of the Old South in the United States. The ideology afflicted not only the slaveowners themselves, but their wives, their mores, the law of the land—and even the self-images of non-slaveowning whites, of overseers, and even the slaves themselves. In one among many passages on this theme, Genovese argues:

This ideology . . . developed in tandem with that self-serving designation of the slaves as a duty and a burden which formed the core of the slaveholders' self-image. Step by step, they reinforced each other as parts of an unfolding proslavery argument that helped mold a special psychology for master as well as for slave. The slaveholders' ideology constituted an authentic world-view in the sense that it developed in accordance with the reality of social relations.
. . . The kind of men and women the slaveholders became, their vision of the slave, and their ultimate traumatic confrontation with the reality of their slaves' consciousness cannot be grasped unless this ideology is treated as an authentic, if disagreeable, manifestation of an increasingly coherent world outlook.[11]

Genovese's marvelously comprehensive account of an earlier capitalist society, where class divisions are obvious, goes into all aspects of the problem—religious legitimations as part of the ideology, and so on. But if his depiction of how economic relations spread out in every direction to become a wholesale ideology seems esoteric and far removed from techno-capitalist ideology, it nonetheless highlights the dialectic of substructure/superstructure in a historian's fashion. The same thing is done from a social-scientific perspective by Peter Berger and Thomas Luckmann.[12] Their focus is on ideological consciousness and how it comes to have the authoritative character it does throughout a culture:

Only at this point does it become possible to speak of a social world at all, in the sense of a comprehensive and given reality confronting the individual in a manner analogous to the reality of the natural world. Only in this way, *as* an objective world, can the social formations be transmitted to a new generation. In the early phases of socialization the child is quite incapable of distinguishing between the objectivity of natural phenomena and the objectivity of social formations. . . . All institutions [including the most basic institution of all, language] appear in the same way, as given, unalterable and self-evident.[13]

It should not be thought that such "objectivity" of social institutions, of ideology, ends when the child grows up. As Berger and Luckmann note, one of the most difficult cases for their dialectical theory of social consciousness is that of the alienated intellectual—and especially of the revolutionary intellectual.[14] But far from disproving the wide-ranging impact of reigning ideologies, the case of the revolutionary intellectual actually confirms the theory: it is extraordinary difficult for *anyone* to break out of an ideology, and, in Berger and Luckmann's view, when one does so, he or she will immediately try to rally a group together and produce a counter-ideology.

Such *praxis*-oriented revolutionary theorizing has been applied directly to technological society and its problems. The best-known instance is the writings of Herbert Marcuse, especially *One-Dimensional Man*.[15] For my part, however, I prefer the elaborations of Marcuse's views, in a historical mode, by David Noble.[16] Where Marcuse claims that any opposition to the reigning ideology—for example, in cases of union opposition to hazards in high-technology industrial workplaces—ends up being interpreted as counterproductive, even irrational (according to the "logical" demands of technological "progress"), Noble spells out in relentless detail, and wherever possible in the words of corporate managers, the *total* way in which the ideology of science and technology in the (alleged) service of society came to permeate every aspect of society in twentieth-century America. To speak of solving particular social problems in our science-based economy without changing the overarching ideology is, in Noble's view, to reinforce rather than undermine the foundations on which the problems rest.

Once again Peter Berger can be cited to provide a social-scientific confirmation of this dialectical view.[17] Berger and his colleagues call their method phenomenological, but they intend for their comprehensive account of how technological production and bureaucracy permeate every aspect of ordinary consciousness in thoroughly "modernized" societies to be taken to be scientific. They believe that it is impossible to conceive of a modern society without technology and bureaucracy, but they are equally convinced that empirical studies will confirm the implications of their account. And to deal with major social problems, such as the boring character of work in highly automated production facilities, in any radical way, without changing the overall technoeconomic system would seem, on their account, to be extremely difficult.

What all of this boils down to is a powerful Marxist objection, that reform politics (sometimes disparaged as "mere procedural justice"[18]) cannot get at the roots of technosocial problems without challenging the technoeconomic system. And that system has built-in disparities between exploiting managers and exploited workers, and between high-technology nations and the so-called "developing nations" so often exploited for the raw materials and exotic minerals needed for high-technology production.

Conclusion? If anything is going to be done to deal with technosocial problems, they cannot be dealt with one at a time. They are all interconnected, and *the* fundamental problem is economic. Only a political revolution that eliminates the power of capitalists and quasi-capitalist bureaucrats over the masses of workers offers any real hope of success.

A PROGRESSIVE LIBERAL RESPONSE

The first issue to deal with here is the claim, often made by Marxists and other radical critics of techno-capitalism, that the liberal response to technosocial problems is "merely procedural"—a technical response depending on administrative law and the courts. Some people who used to call themselves liberals certainly do limit themselves to these approaches; if a technology or environmental impact assessment effort gets coopted by the corporation(s) responsible for the alleged negative impacts, or if the plaintiffs lose in court, they give up, perhaps saying they will return to the fray at the next legal opportunity. However, this need not be the only liberal approach, and some progressive social activists are much bolder than this in their attacks on technosocial problems. I would, indeed, say that the more progressive, socialist-leaning wing of the liberal movement has always been more aggressive than the "merely procedural" objection would suggest. An outstanding example, in my view, is John Dewey. He certainly supported the "polite" legal procedures of the American Civil Liberties Union and the American Association of University Professors (Dewey was a leading member of both), and someone could say that his courageous chairing of the Trotsky Inquiry Commission led to no more than procedural justice for Leon Trotsky; but anyone who reads carefully Gary Bullert's *The Politics of John Dewey*[19] will be convinced quickly that Dewey was a dynamic political activist on a wide range of fronts beyond courts of law. Dewey related many of the social ills he addressed to the rise of technology-based corporations.[20] In a similar way, George Herbert Mead was an activist in local politics in Chicago.[21] I certainly favor procedural reforms, but I am equally convinced that to combat the social evils of technological society much more is needed.

Both Dewey and Mead—and Mead in a special way—were acutely aware of the enormous ideological forces that bear down on an individual who sets out to reform society, and especially a society like ours with the technological propaganda resources at its disposal that Herbert Marcuse has so eloquently pointed out.[22] Nonetheless, though Mead and Dewey were aware of the massive power of a technological-ideological superstructure, neither gave in to technological pessimism. In this, I agree with Mead and Dewey—though there is also a great danger in excessive technological optimism. Such optimism as is called for depends on a belief in the effectiveness of Western-style democracy in dealing with technosocial ills. And this belief, further, rests on a belief—hopefully supported by evidence—that even under pressures of technological propaganda and ideology at least a reform-minded activist minority still has the power to make significant political, social, occupational, educational, and cultural changes.[23]

Given these prefatory notes and beliefs, my principal response to the Marxist objection, above, has three parts.

i. I share Dewey's repugnance to the totalitarian means that have been used by Communist regimes as they have installed their revolutionary forms of government, beginning with the Russian Revolution.[24] As far as I can see, there has been little respect for the civil liberties of ordinary citizens in any Communist country that has sprung up so far. This means that under such regimes, there is little or no opportunity for the sort of public-interest-activist attacks on technosocial evils that I advocate.

ii. There is also, in my opinion, no guarantee that revolution-based Communist regimes, even if they could be established throughout the world and even if they retained their idealistic devotion to the elimination of exploitation, would actually evolve in the direction of a classless society. Marxists, beginning with Marx himself, have always been convinced that once the last vestige of capitalist exploitation is gone—including the elimination of even the possibility of backsliding into capitalist ways by leaders of the new classless workers' society—all exploitation would be gone. A new class of workers will grow up in a non-competitive way, free from all the antisocial impulses associated with capitalist competition, highly productive, capable of exercising creative talents of all sorts, and devoted to the common good. On the other hand, it seems at least conceivable that things would not work out this way at all—that new forms of competition, and eventually of exploitation, would creep back into even the most ideal society. Christians have faith that the community of saints in the afterlife will not return to the sinful ways of this life. Marxists believe that they have discovered scientific laws of historical development that assure the eventuality of a non-regressive, non-exploitative, classless society. Progressive liberals would rather deal with society as we know it, with antisocial as well as socially responsible people and groups dealing with social reforms piecemeal and with no guarantee of total success.

iii. And that brings me to the centerpiece of my progressive-liberal response to the workers' democracy challenge. I believe that progressive social activists are making some headway in terms of combating all nine of the non-economic issues on my list of technosocial ills. Both citizens and governments have begun to respond to pleas of anti-nuclear activists and have taken steps to reduce the grave risks the modern world runs in the proliferation of nuclear arms; there have been notable environmental successes in spite of numerous and continuing setbacks; opponents of genetic engineering and public spying on citizens have not yet made much headway, but at least they are trying to awaken the public to the dangers there; unions and their sympathizers have helped pass some laws on occupational health risks, and others are working to create a climate of concern about boredom and mental health problems associated with highly automated workplaces; social workers do not just do direct-service social work but also get involved in activism to deal with problems of drugs, of stressed-out families in urban centers and so on; more activists are working on technological literacy than on civic preparedness, but that is at least half of the problem in our public school system (though the less important half in my view); and quite a few organized groups are attempting to deal with technological threats to the democratic process in the West. Of my nine issues distinct from technoeconomic (and growing) injustices, that leaves only one item that is not the focus of much social activism—namely, commercialization and corporate domination of the arts.

Given these all-too-limited successes—but, even more, given the wide open future possibilities for public interest activism—I now come to the main point of my reply to the Marxists as well to pessimists who believe in technological determinism. I maintain that if we can continue to make progress in solving these technosocial problems, ours will become a significantly better world. And, even more important, in such an improved climate, there is *at least the possibility that newly empowered citizens will move on from past limited successes to attack head-on the overarching issue of technoeconomic injustices.*

CONCLUSION

For years, what I have advocated as an antidote to the threat of technological determinism (à la Jacques Ellul)—or to the equally serious claim that the threat comes, not from technology simply, but from techno-capitalism—is a revival of the spirit of dissent in public interest activist movements. It may be that, to be really effective, we activists will have to turn from limited successes on particular technosocial problems to form a nationwide or even worldwide New Progressive Movement. And even there we need to pay heed to the warning of people like David Noble. I would take it as a main theme of his *America by Design* that the reform intentions of the old Progressive Movement were distorted by capitalist owners, corporate managers, leading engineers and scientists in industry, educators, and politicians, turning those intentions from true reform to the maintenance of a capitalism threatened by its own inner contradictions. But we have now read Noble and other critics of our social system, and there is no reason in principle why, in our democratic efforts to improve society, we cannot learn from the past. A New Progressive Movement just might be able to tame technology.

I started by saying that, like Marx, I am more interested in improving technological society than in providing a philosophical understanding of it. In my view, Western democracy—if pursued vigorously in the spirit of dissent of the most progressive social activist groups—is more likely to bring about this happy though always improvable state of affairs than any other prescription with which I am familiar.

University of Delaware

NOTES

1. See Paul T. Durbin, "Toward a Social Philosophy of Technology," in P. Durbin, ed., *Research in Philosophy & Technology*, vol. 1 (Greenwich, Conn.: JAI Press, 1978), pp. 67-97, esp. pp. 87 and 97; and "Reviews of Bernard Gendron, *Technology and the Human Condition*, I," in P. Durbin, ed., *Research in Philosophy & Technology*, vol. 3 (Greenwich, Conn.: JAI Press, 1980), pp. 77-87.

2. Willis H. Truitt, "Values in Science," in Durbin, *Research in Philosophy & Technology*, vol. 1, pp. 119-130, esp. pp. 127-130.

3. Bernard Gendron, "Reply: Growth, Power, and the Imperatives of Technology," in Durbin, *Research in Philosophy & Technology*, vol. 3, pp. 102-116, esp. 111-115.

4. Albert Borgmann, *Technology and the Character of Contemporary Life* (Chicago: University of Chicago Press, 1984), pp. 82-85.

5. Manfred Stanley, "Symposium on Albert Borgmann, II. A Critical Appreciation," in P. Durbin, ed., *Technology and Contemporary Life* (Dordrecht: Reidel, 1988), pp. 13-28, esp. pp. 17-21.

6. Stanley (see previous note) lists, as the best example, David Harvey, but also (among others) Louis Althusser, William Connolly, Maurice Godelier, Herbert Guttman, Jürgen Habermas, James O'Connor, E. P. Thompson, and Robert Paul Wolff. See also David McLellan, *Marxism after Marx: An Introduction* (New York: Harper & Row, 1979).
7. See Karl Marx, *Selected Writings*, ed. D. McLellan (Oxford: Oxford University Press, 1977), p. 93.
8. The best recent interpretation is R. W. Sleeper, *The Necessity of Pragmatism: John Dewey's Conception of Philosophy* (New Haven, Conn.: Yale University Press, 1986), where Dewey's philosophy is presented as fundamentally meliorist.
9. In my view, the best recent treatment is Hans Joas, *G. H. Mead: A Contemporary Re-Examination of His Thought* (Cambridge, Mass.: MIT Press, 1985; German original, 1980). Joas draws out the (non-Communist) socialist implications of Mead's social psychology in an admirable fashion.
10. Paul T. Durbin, "Ethics as Social Problem Solving: A Mead-Dewey Approach to Technosocial Problems," in G. Hottois, ed., *Evaluer la technique: Aspects éthiques de la philosophie de la technique* (Paris: Vrin, 1988), pp. 161-173.
11. Eugene D. Genovese, *Roll, Jordan, Roll: The World That the Slaves Made* (New York: Pantheon, 1974), p. 86.
12. Peter Berger and Thomas Luckmann, *The Social Construction of Reality* (New York: Doubleday, 1966).
13. *Ibid.*, p. 59.
14. *Ibid.*, pp. 125-127.
15. Herbert Marcuse, *One-Dimensional Man* (Boston: Beacon, 1964).
16. David Noble, *America by Design: Science, Technology and the Rise of Corporate Capitalism* (New York: Knopf, 1977).
17. Peter Berger, Brigitte Berger, and Hansfried Kellner, *The Homeless Mind: Modernization and Consciousness* (New York: Random House, 1973).
18. Allen E. Buchanan, *Marx and Justice: The Radical Critique of Liberalism* (Totowa, N.J.: Rowman and Allanheld, 1981).
19. Gary Bullert, *The Politics of John Dewey* (Buffalo, N.Y.: Prometheus, 1983).
20. See Larry Hickman, "Doing and Making in a Democracy: Dewey's Experience of Technology," and Edmund Byrne, "Workplace Democracy for Teachers: John Dewey's Contribution," both in P. Durbin, ed., *Philosophy of Technology: Practical, Historical, and Other Perspectives* (Dordrecht: Kluwer, 1989), pp. 97-111 and 81-95.
21. See George Herbert Mead, *Selected Writings*, ed. A. Reck (Indianapolis: Bobbs-Merrill, 1964), pp. xxxiii-xxxvi. Some aspects of this part of Mead's life are available also in Joas's intellectual biography (and interpretation—see note 9, above), but we badly need an adequate biography of Mead.
22. James Campbell, "George Herbert Mead on Social Fusion and the Social Critic," and William M. O'Meara, "Marx and Mead on the Social Nature of Rationality and Freedom." The papers from the Frontiers in American Philosophy Conference, held at Texas A&M University in June 1988, are supposed to be published in a proceedings volume, but details are not yet clear.
23. Michael W. McCann, *Taking Reform Seriously: Perspectives on Public Interest Liberalism* (Ithaca, N.Y.: Cornell University Press, 1986). Morton Schoolman, in "Liberalism's Ambiguous Legacy: Individuality and Technological Constraints," in P. Durbin, ed., *Research in Philosophy & Technology*, vol. 7 (Greenwich, Conn.: JAI Press, 1984), pp. 229-252, bases his optimism about "radicalized liberal politics" on more speculative—but still interesting—foundations.
24. See Bullert's *The Politics of John Dewey* (note 19, above), pp. 130 ff., as well as Dewey's works cited there.

LARRY A. HICKMAN

POPULISM AND THE CULT OF THE EXPERT

One of the most persistent problems of Western industrial societies may be put quite succinctly: "What should be the role of the expert in technologically based democracies?"

This question was the focus of an intense debate that polarized critics of technology in the United States during the 1960s. At one extreme, Emmanuel Mesthene issued a call for massive central planning in which cadres of experts would formulate and implement public policies. At the other extreme, John McDermott offered a solution which merged populism with a mild form of Luddism, eschewed large scale planning, and demanded that the influence of experts be limited. The debate was complicated by the fact that even as Mesthene and McDermott wrote, technocrats employed by the U. S. government were both applying techniques of centralized planning and claiming to march under the banner of laissez-faire capitalism. This was a situation John Kenneth Galbraith appropriately termed "the approved contradiction."

The solutions Mesthene and McDermott proposed were far from novel: they in fact took their place within a longstanding controversy that had been a part of American life since the early decades of this century. As an aid to gaining a perspective on their proposals, I shall examine them alongside the work of John Dewey, who had been a central character in an earlier chapter of the controversy.

When Dewey began to write about technology in the 1880s the cowboy was still the dominant national cultural symbol in the United States, and in France that great monument to the expert, the Eiffel Tower, had not yet been built. By the time of Dewey's 1927 book, *The Public and Its Problems*, however, that Western symbol of lonely, self-reliant individualism had long since been replaced in the public imagination by another hero—the solver of problems of a more public kind, the engineer. As Cecelia Tichi reminds us, the engineer appeared as the hero in over one hundred silent movies and in best-selling novels approaching five million copies in sales between 1897 and 1920. From Edward Bellamy's *Looking Backward* (1888) through *The Education of Henry Adams* (1906) and the writings of Thorstein Veblen, the engineer appears as a crucial figure in modern American civilization. He is in Emerson's sense the representative man for the era, a symbol of efficiency, stability, functionalism, and power. In the imaginative literature of industrialized America he figures as a new hero who enacts the values of civilization. He is at once visionary and pragmatic.¹

Moreover, writes Tichi,

> It is essential to recognize the ways in which engineers and their work were presented to the American public. A masculine figure in a male profession, the spokesman for a gear-and-grinder world, the engineer became a vital American symbol between the 1890s and 1920s. In popular fiction . . . he signified stability in a changing world. He was technology's human face, providing reassurance that the world of gears and girders combined rationality with humanity. And he represented the power of civilization in the contemporary moment. Popular texts present the engineer as the descendant of once-powerful ministers and statesmen; his is ethical *and* utilitarian power. In an industrial era, in a gear and grinder world, the engineer wears the mantle of civilizing power and ethical judgment.²

Herbert Hoover, who would become President of the United States less than twelve months after the publication of Dewey's *The Public and Its Problems*, had a career as an expert—an engineer—and remains to this day the only one of his profession who has been elected to the White House.³

Though not infected by the American tendency to apotheosize the engineer, Dewey was by no means untouched by it. Even as early as the 1890s, long before his extensive articulation of a technologically based version of pragmatism called "instrumentalism,"⁴ Dewey was already using the language of engineering and social planning. His 1891 essay "Moral Theory and Practice" is prescient in its treatment of ethics as a type of engineering problem.⁵ Some decades later, during the difficult period that followed the stock market crash of 1929, Dewey extended his engineering metaphor to social planning of all types. "The first lesson of scientific method," he wrote in 1931,

> is that its fruit is control within the region where the scientific technique operates. Our almost total lack of control in every sphere of social life, international and domestic, is, therefore, sufficient proof that we have not begun to operate scientifically in these fields. . . . For the ultimate issue is not between individualism and socialism, capitalism and communism, but between undisciplined thinking and confused action, and scientific planning and action.⁶

Simply put, Dewey saw the social problems of the Great Depression as consequences of the failure to apply the very techniques that had worked well enough in the sciences and in industry to other public areas in which they might be expected to work quite as successfully. He regarded social problems as complex technical problems.

Dewey's interest extended beyond the engineer proper to a consideration of how all those who are regarded as experts differ from those who are not so regarded. One of his favorite means of explicating his view on this matter was to talk about habits. Taking his cue from William James, Dewey argued that habit was both mainspring and flywheel, both inner tension and live inertial weight, of human action. Habits bind us to old ways of doing things, and they create fear of new courses of action. But the shared habits which we call "institutions" also provide the springboard from which new discoveries can be made and new paths of action entered upon.

Habit and thinking are for Dewey not polar opposites.

> Habit does not preclude the use of thought, but it determines the channels within which it operates. Thinking is secreted in the interstices of habits. The sailor, miner, fisherman and farmer think, but their thoughts fall within the framework of accustomed occupations and relationships. . . . Scientific men, philosophers, literary persons, are not men and women who have so broken the bonds of habits that pure reason and emotion undefiled by use and wont speak through them. They are persons of a specialized infrequent habit.[7]

Dewey's "experts," those who engage their specialized situations intelligently, have learned the methods of utilizing habits in ways that allow greater control of indeterminate and problematic situations than are available to the members of the general public. In contrast to these experts, the majority is caught up in the apparatus of tools and appliances—telephones, automobiles, refrigerators—technological fruits that have come to serve as a surrogate for the critical thought that solves problems. This majority, to whom technological quotidiana have become transparent in their use, stands in stark contrast to the experts—those who have gone beyond satisfaction with the embodiments of technology "in practical affairs, in mechanical devices and in techniques which touch life as it is lived,"[8] that is, beyond preoccupation with the fruits of technology, to the manipulation and control of problematic situations. The expert is one for whom a portion of the technological life world has ceased to be transparent and is instead an object of special focus and considered intervention.

Dewey argued that active engagement of existing habits is effected by means of experimentation, and experimentation is the necessary condition for an understanding of *how* things may be made to work more efficiently. This is no more or less than what Dewey means by "intelligence." Where intelligence is lacking, he suggests, individuals "undergo the consequences, [and] are affected by them. They cannot manage them, though some are fortunate enough—what is commonly called good fortune—to be able to exploit some phase of the process for their own personal profit."[9]

But is there not a danger in all this? Do not the activities of experts, especially when they are directed toward the formulation of public policy, tend to create a new oligarchy, a technocracy which by its very nature is antipathetic to democracy?

Before I examine Dewey's full answer to this question, I shall turn to a different chapter in this long debate. During America's Vietnam era of the 1960s and 1970s, the role of the expert received particularly close scrutiny. Arguments advanced by Mesthene and McDermott during that period are interesting because they offer extremes that Dewey had rejected some forty years earlier.

In *Technological Change: Its Impact on Man and Society*,[10] Mesthene, then the Director of the Harvard University Program on Technology and Society, argued in terms that could at first sight have appeared to be a paraphrase of Dewey's instrumentalism. Technology, he wrote, is more than just hardware. It is "tools in a general sense, including machines, but also including such intellectual tools as computer languages and contemporary analytic and mathematical techniques. That is, we define technology as the organization of knowledge for the achievement of practical purposes."[11] And in language that even more specifically recalls Dewey's, Mesthene summarizes what he says is his book's central argument.

> The argument in brief is that new tools—it takes some of the awe and mystery out of "technology" to think of it as tools—create opportunities to achieve new goals or to do things in new ways. This means that people and groups of people generally must organize themselves differently from before in order to take advantage of the opportunities offered by new tools. (It takes some of the awe and mystery out of the concept of "social institutions" to think of them as groups of people organized in certain ways to accomplish certain purposes.)[12]

Mesthene paints a picture of available tools and methods which lie at the ready, but at the same time in tragic disuse, while vested economic interests and intransigent bureaucracies hold a brake on the application of those tools to persisting social problems. Among the steps that he thinks can be taken to improve this situation are (a) "the proliferation of what might be called scientific or knowledge agencies in government itself,"[13] (b) introduction of the professional scientist into the policy-making process,"[14] and (c) the expansion of "computerized information-handling procedures. . . ."[15]

How will this increased utilization of "professional cadres of technicians and experts" affect traditional democratic forms of social organization? How will people organize themselves differently from before? What kind of new publics will be created? On this point Mesthene articulates a proposal that departs sharply from Dewey's instrumentalist program. His language—which includes talk of tools and publics—clearly reflects his training at Columbia University, where he had received three degrees between 1948 and 1964 and where Dewey's influence was still strong. But his view of the role of the expert in a democratic society could not have been less similar to Dewey's instrumentalist view.

These imperatives of modern decision making . . . run counter to that element of traditional democratic theory that places high value on direct participation in the political process and thus generates the kind of discontent mentioned above. If it turns out on more careful examination that direct participation is becoming less relevant to a society in which the connections between causes and effects are long and often hidden—which is an increasingly "indirect" society, in other words—elaboration of a new democratic ethos and of new political institutions and procedures more adequate to the realities of a modern technological society will emerge as perhaps the major intellectual and political challenge of our time.

At a minimum, it would seem that such new political forms would have to embody a distinction between the expression of preferences and the prediction of consequences. The opportunity to express preferences cannot, in a democracy, be properly denied to any segment of the population. The more complex the society, in fact, and the more technical the issues it has to deal with, the more deliberate must be the effort to allow for the free expression of preferences. . . .

The need for a more refined system of electoral consultation does not however imply that the electorate at large, or any particular segment of it, has a role to play in the technical process of government itself. In a populous, modern, industrialized, and knowledge-oriented society, that process consists increasingly of adumbrating alternative policy options and calculating their probable consequences. It is clearly a job for experts and for all the sophisticated information-handling and management techniques that can be brought to bear on it. To be sure, the policy experts and decision makers must remain accountable to the electorate in the end. . . . But consultation and accountability remain distinct from the technical decision making process, and no amount or combination of them—that is, no amount of "participation" in the populist sense of the term—can substitute for the expertise and decision making technologies that modern government must use.[16]

The role of the public is here reduced to approval or disapproval of policies worked out from afar. In Mesthene's account, democracy is redefined as the ability to recall policy makers, but not necessarily to redirect policy.

A major attack on Mesthene's position appeared in John McDermott's article, "Technology: The Opiate of the Intellectuals," published in *The New York Review of Books* in 1969.[17] McDermott argued that the Harvard report was little more than a call for greater numbers of technocrats who were even further divorced from and impervious to the wishes of the public. He cited as evidence Mesthene's analysis of what he had called "the negative externalities" of technology—pollution, overpopulation, occupational hazards, and the like. Mesthene had suggested that such problems are merely the result of confused signals among the leaders of industry regarding whose business it should be to attend to such matters. McDermott, on the other hand, suggested that the negative externalities had not so much fallen through the cracks as they had been produced by conscious choices made by the captains of industry to put profits above public responsibility. He criticized what he took to be a central claim made by Mesthene and others, among whom he included Zbigniew Brzezinski (who was later to become President Jimmy Carter's National Security Advisor), that the answer to problems caused by technology is just more technology.

> Technology, in their view, is a self-correcting system. Temporary oversight or "negative externalities" will and should be corrected by technological means. Attempts to restrict the free play of technological innovation are, in the nature of the case, self-defeating. Technological innovation exhibits a distinct tendency to work for the general welfare in the long run. *Laissez innover!*[18]

McDermott accused Mesthene and his Harvard group of embracing a modern version of the old laissez-faire doctrines of the empiricists and the utilitarians. "Just as the market of the free play of competition provided in theory the optimum long-run solution for virtually every aspect of virtually every social and economic problem, so too does the free play of technology, according to its writers. Only if technology or innovation (or some other synonym) is allowed the freest possible reign, they believe, will the maximum social good be realized."[19]

Contrary to what he perceived to be Mesthene's "combination of guileless optimism with scientific tough-mindedness," McDermott thought that he saw ample evidence that decisions made by technocrats during the 1960s were in fact undermining democratic politics in the United States. He argued that the decline of popular literacy, the increasing inability of the man or woman on the street to understand the language used by technocrats, was one important sign. He also decried the growth of a two-tiered society which pushed low-level employees into regimented, mind-numbing tasks at the same time that individuals in the upper echelons were able to pursue fulfilling, intellectually challenging work. Finally, he pointed to what he saw as the truncation of the lives of "men and women in the lower and . . . middle levels of American society . . . cut off from those experiences in which near social means and distant social ends are balanced and rebalanced, adjusted, and readjusted."[20]

Much of McDermott's evidence and many of his central metaphors came from the tragic American experience in Vietnam, which he saw as a case study of what goes wrong when technical information is withheld from those who need it in order to do their jobs. It was there that he found incontrovertible evidence of his thesis that the rarified decisions of technocrats inevitably fail when resisted by the experimental methods of those who are concretely involved in specific problematic situations. McDermott argued with considerable irony that the concretely experimental low-tech North Vietnamese and Viet Cong had been more technologically proficient than their high-tech Yankee adversaries.

Although Mesthene and McDermott were ostensibly writing about technology, they were actually writing about different issues under the same rubric. Worse, each writer used the term technology inconsistently. Mesthene employed the word in at least two important senses. The first was an instrumental one: "the organization of knowledge for the achievement of practical purposes." But the second was quite different: technology is whatever is formulated and implemented by experts. Mesthene's unquestioned assumption was that at the level of large social and political organizations, these are the same.

One of Dewey's central arguments in *The Public and Its Problems* had been that technology involves the invention, use, and development of tools of many sorts for the solution of pressing problems. But whenever experts cut themselves off from the concerns of the public, he added, or when the public is excluded from the processes of large scale social planning, then neither the experts nor the public is capable of utilizing the full range of tools otherwise available to them.

McDermott also used the term technology in incompatible senses. Throughout most of his essay he wrote of technology as equivalent to what President Dwight Eisenhower had called "the military-industrial complex." If technology in this sense is said to be self-correcting, then the situation is even worse; because its growth is entrenched within existing power structures, its self-correction can only lead to deeper entrenchment. It is only within a footnote on John Kenneth Galbraith that we finally get a glimpse of McDermott's alternate vision of technology, a vision that was also Dewey's. "My own feeling," McDermott wrote, "is that the fundamental point to take account of in constructing an alternate vision is the fact that technology itself and its need for a skilled and knowledgeable population has created within the population ample resources for self-management even of the most complicated activities."[21]

Both of these opposing arguments acknowledge in passing a view of technology consistent with Dewey's instrumentalism. But then both choose to focus instead on some other view whose drawbacks Dewey had already demonstrated. It is ironic that Mesthene, trained at Columbia, found a technocratic view of technology appealing, or at the very least that he saw no inconsistency between the technocratic view and the democratic one. For his part, McDermott mounted a devastating criticism of Mesthene's technocratic view, but he mentioned his own, more democratic alternative only in passing. It is for this reason that most of his critics have cast him in the role of Luddite.

In brief, Mesthene's essay may be read as a paean to the cult of the expert, and McDermott's as an attempt to tar all forms of expertise with the same unsavory brush. Whereas Mesthene all but ignored the role of the public in policy decisions, McDermott seems unaware of the important role that institutional planners can play in clarifying issues and informing the public. Neither writer attempted to spell out the ways in which constructive interaction between expert and public can be effected.

What were Dewey's suggestions on this matter? In *The Public and Its Problems*, he suggested that dynastic and aristocratic forms of oligarchy had yielded to a new form based on economic considerations. "At all events, it is a shifting, unstable oligarchy, rapidly changing its constituents, who are more or less at the mercy of accidents they cannot control and of technological inventions."[22] What is to be the check on the new oligarchy? Some have argued that ordinary men and women, since they are too distracted to form effective publics, are incapable of providing such oversight. It has therefore been suggested that only an intellectual aristocracy—a cadre of experts of a different sort—would be capable of balancing the interests of experts and public. This scenario pits one group of experts against another for the benefit of the broader public.

Those who hold such a view, Dewey argued,[23] also contend that democratic movements have been just transitional: at one time they provided a necessary antidote to the powers of landed aristocracy and ecclesiastical authority, but they are no longer appropriate to the making of policy in complex technological societies. In short, this view holds that contemporary problems are addressable only by expert intellectuals, even though those experts may in fact form different groups, have different allegiances, and operate from different stances.

This is a strikingly close paraphrase of the conclusion advanced by Mesthene, writing almost exactly four decades later. Against it Dewey mounted a forceful attack. He first argued that if the common people are as "irredeemable" as the argument indicates, then "they at all events have both too many desires and too much power to permit rule by experts."[24] The very factors that allegedly render the common people unfit for rule also make their rule by a class of experts infeasible.

This is a less extreme version of the picture drawn by George Orwell in *1984* of a proletariat perpetually prevented from forming an effective public because of its preoccupation with sex and drugs. In Orwell's novel, the experts found it unnecessary to take steps to control the proletariat, since the proletariat had proved uncontrollable in any case. At the same time, however, the experts had realized that the proletariat had set its own boundaries, constructed its own prison, so to speak, and so constituted no real threat to the continued rule of the experts. The attitude of the experts toward the proletariat was, to use a phrase made famous by Daniel Patrick Moynihan, "benign neglect."

Second, Dewey argued that any attempt by the experts to rule would require either an alliance with the members of the economic oligarchy or else an alliance with non-expert voting classes. In the former case the experts would tend to be subservient to the economically privileged, just as engineers often become the servants of big business. In the later case, ordinary men and women would once again have a share in government.

But Dewey thought that there is an even more serious objection to the cult of the expert, the position eventually endorsed by Mesthene. Experts could never achieve monopoly control over the knowledge required for adequate social planning because "in the degree in which they become a specialized class, they are shut off from knowledge of the needs which they are supposed to serve."[25] In other words, experts can only be experts if they are in close contact with the problematic situations concerning which they have putative knowledge. There is no such thing as an expert in the abstract.[26] This argument is as effective against Plato's vision of an intellectual aristocracy as it is against Mesthene's. In a slightly different form it is also one of McDermott's arguments against Mesthene.

Finally, Dewey followed Alexis de Tocqueville in claiming that a democratic form of government is not just *one* way of associating among many, but the *only* manner of association that is truly educative, both for its officials and for its voting citizens. The strength of democracy is that it forces a recognition that there are common interests, even though the recognition of what they are is confused; and the need it enforces of discussion and publicity brings about some clarification of *what* they are. The man who wears the shoe knows best that it pinches and where it pinches, even if the expert shoemaker is the best judge of how the trouble is to be remedied. Popular government has at least created public spirit even if its success in informing that spirit has not been great.[27]

Dewey here argues that the activation of social intelligence is an affair of knowing in the sense in which knowing is instrumental and experimental—which is to say dynamic, flexible, and in need of constant revision. It was in this context that Dewey rejected the idea that societies should be *planned* and argued instead that they should be *planning*.

But there are many reasons why the expert may fail or refuse to take into account changing public needs: he or she may have put private interests above public ones, or have opted for *a priori* formalism above active experimental knowing. When a class of experts becomes divorced from the public needs they are called upon to serve, then, says Dewey, their knowledge is private knowledge. As far as the public is concerned, this is no knowledge at all.

On the other side, rejecting what would later become the McDermott thesis, Dewey recognized the fact that an uninformed electorate can create prodigious public problems as a result of its failure to vote intelligently, and that it must depend upon experts for the gathering of facts and the construction of scenarios. "Majority rule, just as majority rule, is as foolish as its critics charge it with being."[28] Dewey refused to apotheosize either the decisions of experts, which sometimes tend to be "apart" or *a priori* knowledge, or the popular will, which when implemented without adequate consideration can have disastrous consequences.

Dewey would, for example, have been quite opposed to the proposals put forward by some that public referenda could be expedited by means of electronic voting at home via cable television. What would be missing in such situations would be the debates and deliberations that make citizenship educative and meaningful. At the same time, I think that he would have argued that news programs on commercial television, and even public television's "MacNeil/Lehrer Newshour," fail to live up to their potential. Commercial news programs trivialize information by restricting themselves to headlines and sound-bites, and MacNeil/Lehrer debates tend to present the views of an extremely narrow range of experts who represent almost exclusively the interests of government or industry. A truly democratic news medium would present a much wider spectrum of opinion than is available in either of these venues, with real debates on real issues.

What is important about balloting, Dewey argued, is not the expression of popular will, but the process of antecedent debates, the modification of views to provide adjustments to the opinions of minorities, the discussions, the consultations, and the attempts at persuasion, all of which should precede any public decision that is, to use the vocabulary of William James, "live, forced and momentous."

Dewey thought that the problem of promoting and maintaining meaningful debate in which minority opinions are respected is "*the* problem of the public."[29] Tools, methods, and artifacts are to be measured in terms of the extent to which they promote such interchanges, enhance knowing, and thus facilitate the solution of public problems.

Inquiry, indeed, is a work which devolves upon experts. But their expertness is not shown in framing and executing policies, but in discovering and making known the facts upon which the former depend. They are technical experts in the sense that scientific investigators and artists manifest *expertise*. It is not necessary that the many should have the knowledge and skill to carry on the needed investigations; what is required is that they have the ability to judge of the bearing of the knowledge supplied by others upon common concerns."[30]

Dewey thought that the central challenge of the public lies not in making each of its members an expert, but in reforming and maintaining itself in such a way that enlightened debate concerning its vital interests can continue unimpeded. Against those who would say, as did Mesthene, that the public must in principle be barred from participation in the activities of policy making, and this because of lack of information or sophistication or intelligence, Dewey advanced two counterarguments.

First, such a judgment is based on existing conditions which include the unfortunate effects of secrecy, prejudice, and misrepresentation. There is in fact no way of knowing how capable the public might be if facts were openly available and undistorted by propaganda, and if inquiry were otherwise unimpeded. This was precisely the issue which Dewey faced in defending his view of education against its critics on both left and right. He had no sympathy, for example, toward those who decided upon *a priori* grounds that Irish or Italians or African-Americans were uneducable, as some during Dewey's time thought members of those groups to be.

Dewey advanced a second and related argument. Neither expertise nor the ability to render informed arguments regarding public policy is inborn: "*effective* intelligence is not an original, innate endowment, . . . the actuality of mind is dependent upon the education which social conditions effect."[31] Intelligence in public affairs, like intelligence within all other spheres, is not a private and personal matter, but a function of a rich inheritance of tools and artifacts of all types, including meaning and significance, which has been bequeathed to each individual by his or her forebears. "The notion that intelligence is a personal endowment of personal attainment is the great conceit of the intellectual class, as that of the commercial class is that wealth is something which they personally have wrought and possess."[32]

In sum, the role of the expert in Dewey's vision of a planning society is that of the researcher and clarifier of facts, hypotheses, and the means and methods of bringing about resolution of experienced difficulties. Experts get from members of the public an awareness of what the public problems are and how those problems relate to existing public institutions. They bring to public discourse clarified scenarios and possible courses of action, as well as new insights into pertinent facts. When they fall short of this role or go beyond it, situations arise such as the one described by McDermott in his critique of the policies that led to the disastrous American presence in Vietnam. In such cases the "cadre of experts" described by Mesthene ceases to be a public resource and instead becomes itself a public problem.

Texas A&M University

NOTES

1. Cecelia Tichi, *Shifting Gears: Technology, Literature, Culture in Modernist America* (Chapel Hill: University of North Carolina Press, 1987), p. 98.
2. *Ibid.*, p. 99.
3. Jimmy Carter had some training in nuclear engineering, but was not an engineer by profession.
4. See Dewey's 1903 *Studies in Logical Theory* (*Middle Works* 2:293 ff). References to Dewey's works are to the critical edition, edited by Jo Ann Boydston and published by Southern Illinois University Press. Standard references are to *Early Works* (EW), *Middle Works* (MW) or *Later Works* (LW), followed by volume and page numbers.
5. As late as 1948, Dewey's critics were still pointing to his attempts to treat moral problems as if they were "engineering issues." See Hans Morgenthau, *Scientific Man vs. Power Politics* (Chicago: University of Chicago Press, 1948), p. 36. For more on this matter, see Gary Bullert, *The Politics of John Dewey* (Buffalo, N.Y.: Prometheus Books, 1983), p. 183.
6. Dewey, LW6:50-51
7. Dewey, LW2:335
8. Dewey, LW2:338
9. *Ibid.*
10. Emmanuel G. Mesthene, *Technological Change: Its Impact on Man and Society* (New York: New American Library, 1970).
11. *Ibid.*, p. 25.
12. *Ibid.*, p. vii.
13. *Ibid.*, p. 76.
14. *Ibid.*, p. 77.
15. *Ibid.*
16. *Ibid.*, pp. 79-81.
17. John McDermott, "Technology: The Opiate of the Intellectuals," *The New York Review of Books* (July 31, 1969). There is no anachronism in the fact that McDermott's 1969 article appeared before Mesthene's *Technological Change* in 1970. McDermott's remarks were based on a report published by Mesthene in 1968 which formed the basis for his later book.
18. McDermott, "Technology," p. 99.
19. *Ibid.*
20. *Ibid.*, p. 111.
21. *Ibid.*, p. 121. It is unfortunate that McDermott did not spell out his own alternative more fully. What he seems to have had in mind was a grassroots movement in technology that would initiate bottom-up social planning.
22. Dewey, LW2:362
23. Dewey probably has in mind the views of William Graham Sumner.
24. Dewey, LW2:363
25. Dewey, LW2:364
26. Voltaire had already recognized this fact when he labeled his Dr. Pangloss a "professor of things in general."

27. Dewey, LW2:364
28. Dewey, LW2:365
29. *Ibid.*
30. *Ibid.*
31. Dewey, LW2:366
32. Dewey, LW2:367

JOHN FIELDER

AUTONOMOUS TECHNOLOGY, DEMOCRACY, AND THE NIMBYS

INTRODUCTION

Autonomous technology, in Langdon Winner's view, is "the belief that technology has gotten out of control and follows its own course, independent of human direction."[1] From this standpoint, meeting the demands of technology and its attendant organizations, institutions, and ideology has become the primary goal of society. Instead of technology serving social ends, society has been put in service of technology. Unlike Jacques Ellul, who regards technological dominance of society as virtually total and irreversible, Winner believes that democratic control of technology is possible. At the end of *The Whale and the Reactor*, Winner writes about the Diablo Canyon nuclear reactor, describing it as an intrusive object, an insult to its natural and cultural surroundings. In a public lecture at nearby San Luis Obispo, he proposed that the community should take over the plant and make it a monument to the nuclear age. "For generations thereafter parents could take their children to the place, have leisurely picnics on the beach, and think back to the time when we finally came to our senses."[2]

As if in response to Winner's call, a completed $5.5 billion nuclear reactor at Shoreham, New York, was closed before it had generated a coulomb of electricity. The main issue was the impossibility of evacuating Long Island in case of an accident. (It is frequently impossible for commuters to simply get out of Long Island.) Although there are no plans to officially designate Shoreham as a monument to technological folly, it has already acquired that status in the eyes of its critics. For persons of every ideology, it is a symbol, tragic or hopeful, of the conflict between technology and democracy.[3]

What happened at Shoreham was not unique; it was merely the most recent instance of a nationwide pattern of successful political resistance to the siting of hazardous and otherwise unwanted facilities. At the center of this political phenomenon are the "Nimbys," people who say "Not in my back yard."[4] These are local citizens who organize political and legal opposition to unwanted facilities—like the Shoreham and Diablo Canyon nuclear reactors.

I believe the Nimbys are attempting to bring about democratic limits on technology of the sort Winner calls for. In what follows, I first develop the issue of autonomous technology and democracy, then turn to an examination of the Nimbys and their significance as a movement seeking democratic control over technology. Since the Nimbys have been most successful in opposing hazardous waste treatment facilities, their political significance will be assessed using developments in that area.

AUTONOMOUS TECHNOLOGY

Jacques Ellul's *The Technological Society*[5] is the work that most forcefully introduced the idea of technology as the dominant and dominating feature of contemporary life. More recently, Langdon Winner has examined the concept of autonomous technology, and it is his view that will be used here.

Popular thinking about technology is drawn to its applications: what can we do now that we could not do before? But by focusing exclusively on the immediate applications of technology, we tend to overlook the *requirements* of technology. Institutions, procedures, processes, laws, customs, ideas, and attitudes have been changed and reshaped to accommodate technology. In order to enjoy the products of technology, we have had to create the social, political, and ideological conditions for its development. Government is increasingly oriented toward the business of providing the planning, coordination, and political framework needed for large-scale, technologically based enterprises; social life and institutions either adapt to new technologies or disappear; and much of people's thinking is dominated by technological rationality, which Ellul calls "technique."

The size and power of technologically based industries allow them to exercise extensive control over the government. When private insurance for nuclear reactors was not available, the U.S. government passed the Price-Anderson Act in 1957 which limited the liability of owners in reactor accidents.[6] Except under unusual circumstances, regulatory agencies and their regulations reflect the needs of the industry concerned rather than an independent public interest.[7] Activists in the environmental movement caution against putting too much emphasis on working through oversight agencies because the rules are written by the corporations and government.[8] Given the size of organizations and the amount of investment needed for technology, it is hardly surprising that government has become the servant of technology's needs.[9]

Ellul points to the extraordinary fact that modern societies must continually respond to technological innovations for which they are unprepared. Automobiles led to new forms of living—the suburbs—which changed the nature of neighborhoods, patterns of economic development, and sexual mores of teenagers. Our lack of preparation for these kinds of changes is illustrated by the 1908 British Royal Commission on the Motor Car which considered the most important problem posed by this new technology to be the dust thrown up from unpaved roads.[10] Similarly, television has caused social changes which have largely escaped traditional political scrutiny and control. Society is always several steps behind; technologies are introduced into society and then its institutions, laws, and values must be altered and adjusted to accommodate them. Rather than serving consciously chosen, limited social ends, technology autonomously shapes and reshapes social life according to its needs. Like the sorcerer's apprentice, we have brought technology into being to serve us and instead find ourselves its servants.

The large number and variety of technologies create interactions among these changes which cannot be anticipated and for which there are no existing political mechanisms of control or regulation.[11] The result is an expanding array of institutional adjustments, economic arrangements, and technical requirements without overall direction or control. It is technology determining social reality, outside the control of our democratic institutions.

Ellul holds that the center of autonomous technology is a mode of thought and action, *technique*, which defines problems and possible solutions. In the technological society this mode of thought is dominant, stressing cost-benefit analyses, efficiency, utility, and productivity. In this narrow focus, holistic questions tend to disappear, as do all values that cannot be quantified and analyzed scientifically. Public opinion polls reflect only information that is easily quantifiable and becomes political reality for both candidates and voters. The more narrow concept of feedback replaces the richer notion of response. Computer-scored testing in education tends to reduce education to those portions of students' knowledge and understanding that can be represented by marks in boxes on the answer sheet. And in a perverse form of reverse adaptation, what can be measured by the tests become the purposes of education.[12]

Those on the political front line clearly regard these forms of technological rationality as a trap.

The demand for cost-benefit analysis is an attempt by corporations and their political supporters to define the terrain on which regulatory decisions are made. The complicated mathematical formulas and the arcane terminology exclude ordinary people from the decision-making process. The reliance on scientists and economists makes it appear that regulatory decisions are technical, not political. The use of dollars as the common measure ensures that economic considerations will take precedence over concern for such immeasurable values as health and social justice.[13]

The ideas we use to think about technology often conceal more than they reveal. A crucial distinction in our traditional concept of technology, that of a tool and its use, helps to hide the political dimensions of technology. The distinction holds that tools are value-neutral, able to be used for a variety of purposes. It is only the purposes which have political significance, and the proper response is to discover and punish those who use technology for evil purposes. What is lost in this conception is the array of social and political conditions which must be established in order to make a place for technology.[14] These typically involve important social changes which the tool-use model conceals. For example, the introduction of breeder reactors, which produce weapons-grade plutonium, would also require extensive precautions to insure that plutonium does not fall into the wrong hands. So great is the danger that a commitment to these reactors would likely mean a suspension of many traditional civil liberties for suspects and witnesses in cases of plutonium theft.[15] It is not the *misuse* of breeder reactors that is the problem; it is the way the values and institutions of society must be changed in order to accommodate them.

Autonomous technology thus refers to the pervasive way technology dominates our thinking, social arrangements, and politics. Technology is a way of life, not just an item within life. It requires extensive cultural accommodation as a condition for its existence, which our conventional ways of thinking about technology serve to conceal.

While Ellul is pessimistic about the possibility of change, Winner is more hopeful, calling for a deeper understanding of the politics of technology (going beyond the tool-use model) and creating "a process of technological change disciplined by the political wisdom of democracy."[16] This process would allow other considerations besides technological convenience and economic efficiency to set limits to the social changes we would make. In effect, it would be a new social contract concerning technology. I believe that Winner's analysis is substantially correct and, for that reason, do not foresee democratic changes coming from political institutions that have long been redesigned to serve the needs of technology. The Nimbys, in contrast, are not part of that system. They have initiated a process of discussion and legislative action concerning democratic control of technology, the outcome of which is not yet clear.

DEMOCRACY

Liberal democracy is an attempt to adapt what Benjamin Barber calls "pure democracy" (all of the people governing themselves in all public matters all of the time) to the realities of a modern industrial state. Pure democracy seems impossible in a large, complex society. As a consequence, only some of the people (representatives chosen by election) govern in all public matters all of the time.[17] From the standpoint of autonomous technology, those representatives will be mainly concerned with the needs of technology, so that public policy is shaped by technological momentum rather than other goals. As Ronald Beiner points out, "The challenge for democratic theory is to show that democracy *can* be given a meaningful content in an age where technological imperatives dominate the political agenda."[18]

Other political theorists believe that the primary problem of technology and democracy concerns the citizen's ability to understand technological issues. They point out how difficult it is for ordinary citizens to become knowledgeable about public policy decisions involving technology. H. D. Forbes forcefully states this concern:

Consider some of the issues governments face today: recombinant DNA research, nuclear reactor safety, clean air, ozone depletion, government debt, and the cost of medical care. "What problems like these have in common is that they have enormously important consequences for a vast number of people, they seem to require government decisions of some kind, and in order to make wise decisions, decision-makers need specialized knowledge most citizens do not possess." . . . The democratic process as we know it today—the institutions of liberal, representative democracy—seem incapable of achieving democratic control of the "crucially important and inordinately complex" issues of modern technological society.[19]

The combination of institutionalized technological inertia and citizen ignorance seems to exclude all but a small minority from genuine participation in public policy debates. The forms of democracy remain, but the imperatives of technological society rob them of their substance. Given the severity of these constraints, it is easy to understand Ellul's pessimism about developing limits to technology and hard to be optimistic about Winner's call for us to reassert democratic control.

Barber believes that liberal democracy undermines citizenship, turning citizens into mere voters with only periodic opportunities to exercise their choices on representatives or bond issues. He proposes "strong democracy," in which *all* of the citizens govern in *some* public matters some of the time.[20] Clearly those public matters in which all citizens participate will be matters of great importance to the community.

THE NIMBYS

It turns out that one of the public matters in which local citizens strongly wish to govern themselves is the siting of new facilities of various kinds in their communities. Opposition to unwanted facilities has created a new, noisy, and powerful force in American politics. Local citizens organize, march, sue and petition to block projects in their neighborhoods. Neither powerful capitalists nor marginalized proletarians, Nimbys are generally middle-class and tend to be well-educated, with enough personal and financial resources to organize opposition and hire lawyers. Their strong ties to the community give them a tenacity that is needed for a long fight against the political leverage of organizations seeking a site.

Not all opposition to siting involves technological hazards; Nimbys have blocked group homes for the mentally retarded, McDonald's restaurants, low income housing, and soup kitchens. But Nimbys have been particularly successful in blocking the siting of hazardous facilities. There have been no major hazardous waste dumps sited in the United States since 1980.[21]

Nimbys have predictably been called antitechnology, chemophobic, and selfish, desiring the benefits of technology without its burdens, and foolishly preventing the installation of desperately needed facilities, particularly treatment plants for toxic wastes. Grave warnings have been issued about the impending economic paralysis their actions, if unchecked, will cause. Consultants have been hired to study and pacify them, and a growing body of literature is devoted to the study of ways to resolve siting disputes.[22]

NIMBYS AND DEMOCRACY

"Nimby" originated as a term of abuse, typically defined as "a perverse form of antisocial activism,"[23] and a "social malady."[24] But Nimbys are proud of their designation, replying that "Nimby" is "industry's word for democracy."[25]

Nimbys regard themselves simply as citizens exercising their democratic rights to oppose unwelcome changes in their neighborhood. They rely upon the usual democratic remedies: petitioning their representatives, public meetings and demonstrations, mailings, appearing before local government agencies, particularly those concerned with zoning, and the courts. What is unusual about the Nimbys is the intensity of their opposition and their willingness to use all the democratically available means to oppose an unwanted facility. Rather than simply leave the decision to their elected representatives and government agencies, the traditional pattern in a liberal democracy, Nimbys have acted to make their issue more of a community decision. Nimbys have shown that on siting decisions they believe in "strong democracy," that these decisions should be made by the entire affected community and not just by their representatives. Their success suggests that they have substantial support.

At a time when many thinkers believe that technology is autonomous and out of control, Nimbys seem to have done the impossible: to have reintroduced grass-roots, democratic control over technology. It is as though everywhere, local Davids have arisen to slay technological Goliaths. Those who have successfully fought technological hazards in their neighborhoods speak about a "new people's movement" and a "deeper concept of democracy which gives people a voice in making decisions about their neighborhoods."[26] They, ironically, seem to have reinvented democracy just when some of our best thinkers have declared its disappearance.

One reason that the Nimbys have been successful is that one of the two major obstacles to democratic control of technology raised earlier has turned out to be less important than was previously thought. In an extensive survey of U.S. and European opposition to hazardous waste facilities, Gary Davis and William Colglazier report that,

An uninformed, inexpert public quickly becomes expert when faced with the prospect of a hazardous waste facility. Members of the public may end up knowing more about the potential risks of a given proposal than the regulators or the proponents of the facility.[27]

While it is true that citizens are not well-informed about technologically sophisticated issues being debated in legislatures, they have repeatedly shown that they can become knowledgeable about specific issues that deeply concern them.

Whether even a well-informed community can successfully put limits on autonomous technology is less clear. The Nimbys have brought about some hopeful changes in the way decisions are made about hazardous waste siting, but the significance of those changes is open to debate.

THE NIMBYS AND HAZARDOUS WASTE SITING

In order to understand the Nimbys, it is essential to see their opposition to hazardous facilities siting as not simply an irrational response to the technological needs of society or a childish refusal to accept the price of a way of life that requires the disposal of toxic wastes. Although these responses are certainly present whenever hazardous waste siting is proposed, it is a mistake to see them as the defining characteristic of the Nimbys. The danger—and attraction—of this point of view is that placing Nimbys on the Luddite fringe of society means that they can be easily dismissed as fools, not worthy of serious consideration. The confused and chemophobic pose no intellectual challenge to established ways of thinking about technology and politics.

Roger Kasperson offers a more sympathetic assessment:

Complex technological controversies (such as many risk problems) find publics at a distinct disadvantage, regardless of levels of interest and commitment. Technical experts are divided, risk assessments difficult to understand, value issues poorly defined, and government agencies distrusted. It is not surprising that publics become fearful and risk averse.[28]

The view that Nimbys do not constitute rational opposition to hazardous waste siting permeates much of the writing on this topic. For example, Michael Greenberg and Richard Anderson describe opposition to siting as based on fear and confusion, noting that, "There is every reason to believe that the basis for the public's opinion of hazardous waste sites is not a rational, carefully reasoned consideration of the costs-and-benefits of treating waste at a state-of-the-art facility versus dumping it somewhere."[29]

This is the expert's way of framing the problem. It assumes the framework of technical rationality, focusing on objective, value-free, quantified data expressed in terms of costs and benefits. The judgment of lay people is less valid than that of the experts, so that if Nimbys refuse to accept the expert's view, they are responding to the problem irrationally, opposing siting out of fear and confusion.

Dorothy Nelkin and Michael Pollak call this approach the "welfare model" of public participation in siting decisions. On this view, risks are defined as problems to be solved, mainly by experts, which will result in enhanced institutional legitimacy and social consensus.[30] Public participation is thus conceived as a vehicle to persuade the public to accept controversial technologies.[31] Public participation is simply a manipulative technique used by power-holders, not genuine democratic participation.[32]

The welfare model is essentially paternalistic; citizens are expected to rely on the benevolence of technological elites.[33] Their role in the decision making process is to accept the judgments of experts and ratify their conclusions. The goal is acceptance of controversial siting and restoration of the legitimacy of established decision making institutions and procedures.[34]

TWO CULTURES:
THE TECHNOSPHERE AND THE DEMOSPHERE

It is a mistake to view the Nimbys as persons who fail to meet expert standards for making rational, cost-benefit decisions. Instead, Nimbys and experts can be seen to belong to different cultures, each with its own coherent pattern of ideas, values, and attitudes.[35] One of these, the "technosphere," consists of persons who have been socialized into the culture of technology, the realm of scientific and technical expertise. In contrast are persons in the "demosphere," those whose understandings and actions are rooted primarily in popular culture. In the technosphere, technical rationality reigns, emphasizing measurement, rigorous methodology, and avoidance of explicit value positions. Those in the demosphere show considerable skepticism about claims of scientific rationality, stressing the need to humanize, subjectify, and contextualize key issues.[36] Thus it is not surprising that technologists talk about their cost-benefit analyses while affected citizens talk about a particular neighborhood and what it means to them.

Tod Crumrine is fighting a huge garbage incinerator proposed for a site a thousand feet away from his farm. He is the fifth generation of his family to live on his farm in rural Nova, Ohio.[37] His long-standing tie to the land just does not fit into the calculations of technological rationality. It concerns the value of what is personal, historical, and individual against what is efficient, quantifiable, and utilitarian. What is real for the incinerator company is the suitability and economic value of the site; Tod Crumrine's deep attachment to *this* piece of land simply does not count in the technosphere any more than the disharmony of reactor and canyon counted in California.[38] His attempt to block the incinerator is more than an attempt to avoid having a garbage incinerator for a neighbor; it is also the fight to have the value of his historic attachment to the land included in the decision making process. From the standpoint of the technosphere this is incomprehensible, irrelevant, irrational, or just unfortunate, but to people like Tod Crumrine, who have this kind of relationship to a particular community, nothing is more real and worth fighting for.[39]

The two cultures are clearly reflected in the argument, often made by facilities proponents, that if people really understood what the experts know they would no longer oppose siting. Harry Otway quotes an American industrialist: "People need to be better educated about our industry and the techniques we use. If they had the same knowledge we have, they would see things our way and we could reduce uncertainty and costly delays." Otway asks if the converse is also true: would the industrialist oppose siting if he had the same knowledge as lay people about how his technology would affect their lives?[40] Each imagines that if the other could see the world from his point of view, there would be agreement. But, of course, the reason they do not is that each sees the problem at hand from a much different framework of values and beliefs.

Nimbys assert the values of the demosphere against those of the technosphere. Their opposition directly challenges the exclusive claim to validity of technological rationality, which is an essential ingredient of autonomous technology. Democratic control of technology places the concerns of the *demos* over the demands of *techne*. It is an intellectual claim that calls technological rationality itself into question and demands that limits be set according to the priorities of the demosphere.

EVOLUTION OF SITING POLICIES

It is not surprising that attempts to merely placate the Nimbys were unsuccessful. The issue is not technical information understood only by experts, but who exercises political control over the siting and operation of hazardous facilities. It is a political issue and requires a political solution. Nimbys rejected the assurances of the experts and refused to play their assigned role in the welfare model. As a result there has been a steady development of new policy proposals addressing the siting problem and the role of the public.

The public outcry over toxic waste, particularly hidden waste sites like Love Canal, led to a flurry of political and legislative activity. The initial response of public officials was to create an appropriate mix of legislation, agency control, and market incentives that could induce private industry to incorporate waste disposal into their thinking and planning. Thus, the problem would be solved mainly through private initiatives.[41] But intense public opposition to new toxic waste sites in conjunction with other factors undermined this approach. Nimbys, aided by marketplace uncertainty, huge liability risks, and complicated permitting requirements, made siting facilities impossible in many localities throughout the United States.

Different versions of the conventional policy approach were tried. By 1987, thirty-six states had shifted the locus of siting decisions from localities to the state government. In some cases the aim was simply to pre-empt local opposition,[42] but in most cases the move simply reflected the failure of local government and industry to solve the problem. Many observers hoped that at the state level it would be politically easier to balance competing interests and provide offsetting compensation to affected communities. However, since the specifics of oversight and regulation were still in local hands—health and fire departments—"the bottom line had not changed."[43] Nimbys were still able to prevent siting.

These approaches remained firmly within the welfare model. In this strategy, public participation is shifted to a more remote, indirect, and less accessible level of government, and the decision is still a matter for government and industry experts. Hence this phase of dealing with the Nimbys simply sought a more successful locus of decision making within traditional political forms.

NEW FORMS OF PUBLIC PARTICIPATION

Faced with continued siting paralysis, conventional assumptions about the relations between government, business, and citizens are currently being reexamined and changed. What is happening is the development of new forms of public participation and decision making that more fully involve the persons affected.[44]

The traditional siting process is a matter of "decide, announce, and defend."[45] A developer acquires a site, announces the decision, and applies for the needed local, state, and federal permits. Individuals and groups may use public hearings and litigation to express their views. Effective participation requires the use of lawyers, and the entire process is conducted on an adversarial basis.[46] Law, custom and procedural rules reflect this approach.

New formal mechanisms which emphasize negotiation, face-to-face discussion and mediation are now being discussed and developed.[47] A procedure that draws upon this work is embodied in the Massachusetts Hazardous Waste Siting Act of 1980.[48] It requires that builders of hazardous waste facilities must negotiate with the affected communities concerning compensation for increased costs or risks associated with the facility. The law mandates the creation of local assessment committees to represent the affected communities.

The results of negotiations between builder and community in siting decisions are embodied in a contract between the owner-operators and the community.[49] Such contracts would be legally enforceable documents specifying exactly what each party would do. Contracts would include items such as compensation to the community for inconvenience and other losses, requirements for risk mitigation and how adverse impacts, like accidents, would be handled. Another important item that the contract could include is a performance agreement which sets out timetables and targets for reduction of various types of emissions, source reduction, and any other feature of the operation which concerns the community.[50]

Facilities contracts are flexible enough to meet the individual circumstances of a site, the community, and the owner-operator. Instead of relying on notoriously weak government permitting and enforcement, with its minuscule fines and tendency to allow facilities to operate "temporarily" in violation of its permit, the judicial system becomes the instrument of enforcement through breach-of-contract proceedings.[51] This gives the community more direct control than the present system.

Facilities contracts may include collaborative community-management plant oversight to maintain the role of the community after the plant is in operation. Members of the community, like Michael Elliott of the Georgia Institute of Technology, directly participate in the guidance of the operation, serving as the community's representative in management.[52] A more radical step which has been successful in Europe is a partnership between the company and the community in both the ownership and operation of the facility.[53] Such arrangements would require a further step away from our traditional separation of public and private activities.[54] That these proposals are now being considered and enacted represents a willingness to revise our traditional ideology of technology in the face of pressure to solve siting problems.

CRITIQUE OF NEGOTIATION AND MEDIATION

Although negotiation between builder and community is better than the traditional procedure of "decide, announce, and defend," there are reasons to believe that, in many cases, it is simply a more elaborately disguised version of the welfare model of siting. Douglas Amy notes that mediation is only appropriate in certain political conditions, particularly where there is a rough balance of power between the disputants. That balance, he claims, has decisively shifted against environmental groups in the last fifteen years because of increased spending and organization by industry, a change in attitude toward environmental issues at the Federal level, and setbacks for environmentalists in the courts. Under these circumstances, mediation can easily produce inequitable and cooptive agreements, particularly where mediation is legally required, as in the Massachusetts Hazardous Waste Facility Siting Act.[55]

Successful mediation requires agreement about common objectives and values; without these it becomes simply a power struggle. But one of the characteristics of siting disputes is the lack of agreement on fundamentals, so much so that the conflict can be characterized as a clash of two cultures, the technosphere and the demosphere. And Nimbys have become increasingly concerned not only with the site in their back yard, but also with the broader political and social assumptions that underlie siting.

FROM NIMBY TO NIABY

The Not in My Back Yard movement has evolved along with proposals for resolving siting disputes. Opponents of different hazardous waste sites have joined together to say, "Not in *anybody's* back yard."[56] The essence of the Niaby approach is to challenge the issue of waste disposal as a *siting* problem. Finding a place to put hazardous waste assumes the validity of rear-end disposal and of framing the problem in terms of a production system in which producers generate large amounts of waste which are then disposed of in separate sites. Besides causing battles with local residents, costly disposal facilities require a large, continuous waste stream to operate efficiently and thus compete with source reduction and recycling efforts. Defining hazardous waste as a siting problem accepts the current production and disposal system, which the Niabys reject.

Not only does the welfare model of the technosphere make important assumptions about how siting decisions are to be made and what values are to be used, but it also makes important political and social assumptions about production, consumption, and disposal. From the welfare perspective, hazardous waste sites are socially desirable, a public good, and the issue is how best to site them. Niabys do not see them as a public good but as a symptom of a defective system of production that is destroying the environment and the quality of life. They call for a commitment not to finding new sites but to a program of source reduction, outright bans on some synthetic chemicals and toxic metals, and on-site destruction of toxic wastes.[57] Niabys seek political control not simply of siting but of the economic and technological apparatus that produces hazardous wastes in the first place.

THE SIGNIFICANCE OF THE NIABYS

In recent years, Nimbys/Niabys have exercised a virtual veto power over hazardous waste siting. Their initially local concerns have evolved into a political critique of the system of production and disposal of hazardous wastes in society at large. Their movement has gone from isolated groups to the formation of umbrella organizations to coordinate environmental issues all over the United States.[58]

Niabys have become increasingly sophisticated in their use of the political process and in their understanding of the political and economic dimensions of siting. They have evolved from independent, local groups to a coordinated movement with shared goals and information. At the same time they have, to an increasing extent, directed their attacks at the structural basis of the problem and have demanded changes that would preserve the values of the demosphere.

They have made an impact not primarily through the institutions of representative democracy, which they regard as captives of technological interests, but through local activism, both political and legal. It is the classic approach of those outside the political power centers, and the Niabys have many parallels with the civil rights movement, the Vietnam war protests, and the 1989 student uprising in Beijing. It is "the political wisdom of democracy" asserting itself in a political system that is powerfully biased in favor of runaway technology and whose social, economic, and political arrangements are designed to support it. It is not surprising that democratic control over technology appears not as a mainstream political issue but as an almost revolutionary challenge to the existing order.

The concern over democratic control of technology has focused on the captive institutions of *representative* democracy and overlooks the strength of democratic action made possible by constitutional guarantees of freedom of press, assembly, and legal redress, and by a political system that allows significant local control over zoning, health, and fire regulations.

The Niabys have been successful largely because of two factors. First, in hazardous waste siting the price of our technological society becomes subjectively real and directly threatens community values. Grassroots movements do not arise in response to distant, abstract problems but in response to clear and present dangers to the community. The second important factor is local control over zoning, health, and fire regulations. This gives Niabys a direct and powerful way to challenge the industrial and government interests that would easily defeat them in the state and federal agencies. With these two conditions present, democratic control over technology seems a real possibility.

The democratic limits on autonomous technology imposed by the Nimbys/Niabys are fragile. They depend heavily on the ability to stop or greatly delay siting by local control over site characteristics. The effectiveness of opposition to hazardous waste sites and nuclear power facilities depends upon the accessibility of the government. In France and Japan, which "are among the least open to popular influence of any of the world's representative systems of government," citizen opposition has been much less effective.[59]

Removing control to the state level has been the tactic of choice to circumvent the Nimbys. While this has not been successful so far, there are no guarantees that it will not ultimately prevail. The nascent Niaby movement is not very strong, and, as noted, Amy points out that the environmental movement has lost strength relative to industry in the last fifteen years.

Niabys see themselves as part of a movement against the domination of life by autonomous technology and the large organizations associated with it. They are dedicated to establishing the limits to technology that Winner called for. Perhaps in a few years "Nimby" will no longer be a term of abuse, and when we use it we will "think back to the time when we came to our senses."

Villanova University

NOTES

1. Langdon Winner, *Autonomous Technology* (Cambridge, Mass.: MIT Press, 1977), p. 13.
2. Langdon Winner, *The Whale and the Reactor* (Chicago: University of Chicago Press, 1986), p. 177.
3. Philip S. Gutis, "Shoreham Saga Comes to a Bitter End," *New York Times* (May 29, 1988), Week in Review Section, p. 5.
4. William Glaberson, "Coping in the Age of 'Nimby'," *The New York Times* (June 19, 1988), section 3, pp. 1 and 25. See also A. Amour, ed., *The Not-in-My-Backyard*

Syndrome (Downsview, Ontario: York University Press, 1984), and Bryant Wedge, "The NIMBY Complex: Some Psychological Considerations," in Institute for Environmental Negotiation, *Not-in-My-Backyard: Community Reaction to Locally Unwanted Land Use* (Charlottesville, Va: University of Virginia, Institute for Environmental Negotiation, 1984).

5. Jacques Ellul, *The Technological Society*, trans. John Wilkinson (New York: Knopf, 1964). See also *The Technological System* (New York: Continuum, 1980).

6. Joseph G. Morone and Edward J. Woodhouse, *The Demise of Nuclear Energy?* (New Haven, Conn.: Yale University Press, 1989).

7. See Grant McConnell, *Private Power and American Democracy* (New York: Knopf, 1966); Ellul, *Autonomous Technology*, pp. 243-244; Harry Otway, "Experts, Risk Communication, and Democracy," *Risk Analysis* 7:2 (1987): 125.

8. Nicholas Freudenberg, *Not in Our Backyards!: Community Action for Health and the Environment* (New York: Monthly Review Press, 1984), pp. 257-258. Similar sentiments are expressed concerning legal remedies (pp. 168-169). Nimbys have been most successful in organizing opposition at the local level.

9. One of the best statements of this relationship is in John Kenneth Galbraith, *The New Industrial State*, 3d ed. (New York: New American Library, 1979).

10. David Collingridge, *The Social Control of Technology* (New York: St. Martin's, 1984), p. 16.

11. Winner, *Autonomous Technology*, p. 89.

12. *Ibid.*, pp. 215, 223, 235.

13. Freudenberg, *Not in Our Backyards!*, p. 81.

14. See Winner, *Autonomous Technology*; also, *Whale and Reactor*, "Technologies as Forms of Life," pp. 3-18.

15. This example is discussed in Winner, *Whale and Reactor*, p. 37.

16. *Ibid.*, p. 55.

17. Benjamin Barber, *Strong Democracy* (Berkeley: University of California Press, 1984), p. xiv.

18. Richard B. Day, Ronald Beiner, and Joseph Masciulli, eds., *Democratic Theory and Technological Society* (Armonk, N.Y.: Sharpe, 1988), p. xi.

19. Quoted by Ronald Beiner in *Democratic Theory and Technological Society*, p. xi.

20. Barber, *Strong Democracy*, chapter 6.

21. *Time* (June 27, 1988), p. 45.

22. The literature on this topic is substantial. The following books are a good place to enter it. Douglas J. Amy, *The Politics of Environmental Mediation* (New York: Columbia University Press, 1987); Gail Bingham, *Resolving Environmental Disputes* (Washington, D.C.; The Conservation Foundation, 1986); Christopher J. Duerksen, *Environmental Regulation of Industrial Plant Siting* (Washington, D.C.: The Conservation Foundation, 1983); David Morrell, *Siting Hazardous Facilities* (New York: Ballinger, 1982); Michael O'Hare, Lawrence Bacow, and Debra Sanderson, *Facility Siting and Public Opposition* (New York: Van Nostrand Reinhold, 1983; Timothy J. Sullivan, *Resolving Development Disputes through Negotiation* (New York: Plenum Press, 1984).

23. *Time* (June 27, 1988), p. 44.

24. Kent E. Portney, "The Dilemma of Democracy in State and Local Environmental Regulation: The Case of Citizen Participation in Local Hazardous Waste Treatment Facility Siting," a paper prepared for delivery at the 1987 annual meetings of the Midwest Political Science Association, Chicago, Ill., panel 7-4a, April 10, 1987, p. 1.

25. *Everyone's Back Yard*, 7:1 (Spring, 1989): 6. Published by the Citizen's Clearinghouse for Hazardous Wastes, Inc., P.O. Box 926, Arlington, VA 22216.

26. Freudenberg, *Not in Our Backyards!*, pp. 9 and 41.

27. Gary Davis and Bruce Piasecki, "Siting Hazardous Waste Facilities: Asking the Right Questions," in Bruce W. Piasecki and Gary A. Davis, *America's Future in Toxic Waste Management* (New York: Quorum Books, 1987), p. 174.

28. Roger E. Kasperson, "Six Propositions on Public Participation and Their Relevance for Risk Communication," *Risk Analysis* 6:3 (1986): 278. Similar sentiments can be found in O'Hare, Bacow, and Sanderson, *Facility Siting and Public Opposition*, pp. 2-3.
29. Michael R. Greenberg and Richard F. Anderson, *Hazardous Waste Sites: The Credibility Gap* (New Brunswick, N.J.: Center for Urban Policy Research, 1984), p. 288.
30. Dorothy Nelkin and Michael Pollak, "Public Participation in Technological Decisions: Reality or Grand Illusion?" *Technology Review* (August/September, 1979), pp. 55-64.
31. *Ibid.*, p. 55.
32. Harry Otway, "Experts, Risk Communication, and Democracy," *Risk Analysis* 7:2 (1987): 127.
33. A good discussion of this point is in Leslie Sklair, "Science, Technocracy, and Democracy," in D. Elliott and R. Elliott, eds., *The Politics of Technology* (London: Longman, 1977), pp. 172-185.
34. Nelkin and Pollak, "Public Participation in Technological Decisions," p. 55.
35. This is the approach of John D. Powell and Kent E. Portney, "Normative Conflicts in Environmental Politics and in the Analysis of Environmental Policy," a paper prepared for a panel at the 1988 meetings of the Western Political Science Association, San Francisco, Calif., March 12, 1988, p. 2.
36. *Ibid.*, pp. 3-4. A more extended discussion is in Mary Douglas and Aaron Wildavsky, *Risk and Culture* (Los Angeles, Calif.: University of California Press, 1982).
37. *New York Times* (June 19, 1988), section 3, p. 1, 25.
38. A similar process took place with the displacement of the American Indian. Apart from the question of the economic value of the lands given to replace their original homeland, whites simply disregarded the attachment of tribes to specific, individual places. Their lands had meaning to them in the same way Tod Crumrine is attached to his particular piece of Ohio.
39. A lively philosophical justification for cultural, ethical, and political concerns of this kind can be found in Mark Sagoff, *The Economy of the Earth: Philosophy, Law and the Environment* (New York: Cambridge University Press, 1988).
40. Otway, "Experts, Risk Communication, and Democracy," p. 127.
41. Daniel Mazmanian and David Morell, "The Elusive Pursuit of Toxics Management," *The Public Interest* 90 (Winter 1988): 86.
42. Portney, "The Dilemma of Democracy," p. 2.
43. Mazmanian and Morell, "The Elusive Pursuit of Toxics Management," p. 91. See also Richard N. L. Andrews and Terence K. Pierson, "Local Control or State Override: Experiences and Lessons to Date," *Policy Studies Journal* 14:1 (September 1985): 90-99.
44. Mazmanian and Morell, "The Elusive Pursuit of Toxics Management," p. 92.
45. Kasperson, "Six Propositions on Public Participation," p. 277.
46. Bingham, *Resolving Environmental Disputes*, p. 201.
47. *Ibid.*, p. 5.
48. The act is reprinted in O'Hare, Bacow, and Sanderson, *Facility Siting*, pp. 192-214. For a discussion of the act, see pp. 182-192.
49. Mazmanian and Morell, "The Elusive Pursuit of Toxics Management," p. 94; O'Hare, Bacow, and Sanderson, *Facility Siting*, p. 169.
50. Mazmanian and Morell, "The Elusive Pursuit of Toxics Management," pp. 93-94.
51. *Ibid.*, p. 94.
52. *Ibid.*, p. 95.
53. Some cases are discussed in Nelkin and Pollak, "Public Participation in Technological Decisions"; see also Meinolf Dierkes, Sam Edwards, and Rob Coppock,

eds., *Technological Risk: Its Perception and Handling in the European Community* (Cambridge, Mass.: Oelgeschlager, Gunn and Hain, 1980).

54. See Michael L. P. Elliott, "Improving Community Acceptance of Hazardous Waste Facilities through Alternative Systems of Mitigating and Managing Risk," *Hazardous Waste* 1:3 (Fall 1984): 404.

55. Amy, *Politics of Environmental Mediation*, pp. 200-212.

56. One of the major organizations of Niabys, the Citizen's Clearinghouse for Hazardous Waste, Inc., publishes a newsletter; see note 25 above.

57. This description of the Niabys is taken from Michael Heiman's "Not in Anybody's Backyard: The Grassroots Challenge to Hazardous Waste Facility Location," a paper presented at the Association of American Geographers annual meeting in Baltimore, Md., March 19-22, 1989.

58. An advertisement on page 33 in the September, 1989, *Harper's* magazine features Lois Gibbs, Executive Director of the Citizen's Clearinghouse for Hazardous Waste. She endorses Working Assets VISA, which supports peace, human rights, and environmental groups.

59. Morone and Woodhouse, *The Demise of Nuclear Energy?*, p. 35.

KRISTIN SHRADER-FRECHETTE

TECHNOLOGY, BAYESIAN POLICYMAKING, AND DEMOCRATIC PROCESS

Contemporary assessments of technological risks are fundamentally anti-democratic in at least one sense. They typically employ decision rules that take little or no account of the hazard evaluations of laypersons most likely to be harmed by a risky technology. Instead, assessors usually assume that experts, rather than the people, have the right to define which risks are rationally acceptable.

Although there are many ways in which technology assessment often disenfranchises the public, use of Bayesian rules for determining rational behavior under risk is one of the most important. This essay will argue that one particular Bayesian rule is anti-democratic in at least two important senses. (1) It rejects *egalitarian* in favor of *utilitarian* norms for making policy regarding risky technology. (2) It presupposes that the *outcome* of various decisions regarding hazardous technology, rather than the democratic *process* by which they are made, legitimates them.

1. THE IMPORTANCE OF RULES REGARDING RATIONAL RISK BEHAVIOR

Were certain Bayesian rules for technological decisionmaking rejected or amended, it can be shown not only that democracy would benefit, but also that many controversial policies regarding acceptable technology would be reversed. Assessments of nuclear technology provide a prominent example of this point. The controversial government report, WASH-1400, calculated the per-year, per-reactor probability of a nuclear core melt as 1 in 17,000; for the 151 existing and planned U.S. reactors, the probability of a core melt during their lifetime comes to one in four.[1] Although both of these numbers seem uncomfortably large, risk assessments done by both the Ford Foundation-Mitre Corporation and by the Union of Concerned Scientists (UCS) agree that the WASH-1400 probabilities were too small by approximately an order of magnitude.[2] Paradoxically, although both these latter studies agree on the probability and consequence estimates associated with the risk from commercial nuclear fission, they make diametrically opposed recommendations regarding the advisability of using atomic energy to generate electricity. The UCS risk analysis decided against use of the technology; the Ford-Mitre study advised in favor of it.[3]

How could hazard assessments that agree on the probability and consequences associated with a particular technological accident reach contradictory conclusions about the advisability of using the technology? In many such cases, including that of the UCS-Ford/Mitre controversy, the reason is that the two studies used quite different methodological rules for determining rational behavior under risk, quite different rules at the third, risk-evaluation stage of assessment. The Ford-Mitre research was based on the widely accepted Bayesian decision criterion that it is rational to choose the action with the best expected value or utility. "Expected value" or "expected utility" is defined as the weighted sum of all possible consequences of the action, and the weights are given by the probability associated with each consequence. The UCS recommendation followed the decision rule that it is rational to choose the action that avoids the worst possible consequence of all options.[4]

To know whether a particular policy analysis provides the best risk evaluation, the best recommendation for rational behavior regarding risk, we obviously need to know under what circumstances a particular decision rule is to be preferred. Ought we to be technocratic utilitarians and choose a Bayesian rule? Or ought we to be cautious egalitarians and follow a maximin strategy of preferring the action whose worst consequence is less serious than the worst consequence of all possible actions?[5] The "prevailing opinion" among scholars, according to John Harsanyi, is to use the Bayesian rule.[6] As he puts it: "In the case of risk, acceptance of Bayesian theory is now virtually unanimous. In the case of uncertainty, the Bayesian approach is still somewhat controversial, though the last two decades have produced a clear trend toward its growing acceptance by expert opinion."[7] The prevailing opinion among experts is that one ought to dismiss the non-Bayesian decision rules of laypeople as egalitarian, pessimistic, and irrational,[8] and accept the Bayesian rules as utilitarian, optimistic, and rational. Contemporary risk assessors err, I shall argue, because expected utility is not the most desirable decision rule, especially in situations of uncertainty regarding technological risk. Even though it dominates policymaking, such a rule ought not be used in circumstances characterized by the possibility of low-probability, high-consequence events, e.g., a nuclear core melt. I shall argue that, in such cases, an anti-democratic rule could cause grave harm.

2. HARSANYI VERSUS RAWLS: EXPECTED UTILITY VERSUS MAXIMIN

Perhaps the most famous contemporary debate over which decision rules ought to be followed in situations of technological risk and probabilistic uncertainty is that between Harvard philosopher John Rawls and Berkeley economist Harsanyi. Harsanyi takes the Bayesian position that we ought to employ expected-utility maximization as the decision rule in situations of uncertainty, as well as in those of certainty and risk.[9] Typical conditions of *uncertainty* are when we have partial or total ignorance about whether a choice will result in a given outcome with a specific probability, e.g., when we have partial ignorance about whether a choice of using nuclear power to generate electricity will result in the probability that at least 150,000 people will die in a core-melt accident. Under conditions of *certainty*, we know that a choice will result in a given outcome, e.g., if we use nuclear power to generate electricity, then we are certain to have the outcome that there will be radwaste to manage. Choices between bets on fair coins are classical examples of decisions under *risk*, since we can say that we know, with a specific probability, whether a choice will result in a given outcome. (Let us call the Bayesian sense of risk, involving specific, known probabilities, "$risk_B$" and the sense of technological risk, often involving uncertainty, with which hazard assessors deal, simply "risk.")

Most technology-related decisionmaking probably takes place in situations of uncertainty. We rarely have complete, accurate knowledge of all the probabilities associated with various outcomes of taking technological risks, e.g., from hazards such as pesticides, liquefied natural gas facilities, and toxic wastes. In part this is because the very risky technologies that most need assessing are often new; since they are new, we have insufficient empirical data upon which to establish a frequency record for various accidents. The U.S. National Academy of Sciences has been so concerned with this problem that in 1983 it undertook a study of the use of risk assessment in the U.S. government to examine why it had often become the focus of debate instead of alleviating regulatory controversies. The report concluded that the basic problem is the "sparseness and uncertainty of the scientific knowledge of the health hazards addressed."[10] This means that, in the absence of complete and accurate probabilistic knowledge, most of the pressing problems of risk analysis concern uncertainty, not "$risk_B$." But if so, then one of the most important questions in the Harsanyi-Rawls, expected utility-maximin debate is what decision rule ought to be followed under the conditions of uncertainty that characterize the state of our knowledge about some of our most beneficial and controversial technologies. This is the question we shall address.

Harsanyi believes that under conditions of uncertainty, we should maximize expected utility, where the expected utility of an act for a two-state problem is

$$u_1 p + u_2(1-p)$$

where u_1 and u_2 are outcome utilities, where p is the probability of S_1 and (1-p) is the probability of S_2, and where p represents the decisionmaker's own subjective probability estimate.[11] More generally, members of the dominant Bayesian school claim that expected-utility maximization is the appropriate decision rule under uncertainty.[12] They claim that we should value outcomes, or societies, in terms of the average amounts of utility (subjective determinations of welfare) realized in them. To maximize expected utility in societal choices, Harsanyi opts for using the maximum average utility level of all individuals in society.[13]

Proponents of the maximin school maintain that one ought to maximize the minimum, i.e., avoid the policy having the worst possible consequences.[14] Many of them, including Rawls, take the maximin principle as equivalent to the *difference principle*. According to a very simplififed version of this principle, one society is better than another if the worst-off members of the former do better than the worst-off in the latter.[15]

In terms of social choice, the obvious problem is that often the maximin and the Bayesian/utilitarian principles recommend different actions. To consider how this can happen, consider an easy case.

Imagine two societies. The first consists of a thousand people, with a hundred being workers who are exposed to numerous occupational risks and the rest being free to do whatever they wish. We can assume that, because of technology, the workers are easily able to provide for the needs of the rest of society. Also assume that the workers are miserable and unhappy, in part because of the work and in part because of the great risks that they face. Likewise, assume that the rest of society is quite happy, in part because they are free not to work and in part because they face none of the great occupational risks faced by the hundred workers. This means, of course, that the compassion of the nine hundred nonworkers does not induce them to feel sorry for the workers and to feel guilty for having a better life. Hence we must assume that the nonworkers' happiness is not disturbed by any feeling of responsiblity for the workers. We must assume that the nonworkers have been able to convince themselves that each of the workers and their children were given good educations and equal opportunity. Likewise we must assume that the nonworkers believe that the workers were able to compete for the positions of nonworkers, and that, since the workers did not try hard enough, and work diligently enough to better themselves, therefore they deserve their state. With all these (perhaps implausible) assumptions in mind, let us suppose that, using a utility scale of one to a hundred, the workers each receive one unit of utility, whereas the others in society each receive ninety units each. Thus the average utility in this first society is 81.1.

Now consider a second society, similar to the first, but in which, under some reasonable rotation scheme, everyone takes a turn at being a worker. In this society everyone has a utility of thirty-five units. Bayesian utilitarians would count the first society as more just and rational, whereas proponents of maximin and the difference principle would count the second society as more just and rational.[16]

Although this simplistic example is meant merely to illustrate how proponents of Bayesian utilitarianism and maximin egalitarianism would sanction different social decisions, its specific assumptions make maximin in this case appear the more reasonable position. Often, however, the reverse is true. There are other instances in which the Bayesian position is obviously superior, especially in situations of $risk_B$ or certainty. In this essay, we shall be concerned merely with an appropriate decision rule for cases of *uncertainty* (not $risk_B$), since technological hazards are often typified by uncertainty.

Obviously the more general question of using Bayesian versus maximin decision rules, in all cases, is too broad to be settled here. This is in part because it mirrors a larger controversy in post-medieval moral philosophy between the utilitarians and the contractarians. The utilitarian tradition is represented by David Hume, Adam Smith, Jeremy Bentham, John Stuart Mill, and many contemporary social scientists and moral philosophers such as Harsanyi or J. J. C. Smart. The contractarian (or social-contract) tradition is represented by John Locke, J.-J. Rousseau, Immanuel Kant, and by contemporary moral philosophers such as Rawls. The analysis in this essay is directed at questions of *societal* risk-taking, because of their importance to democracy. It is not directed at the larger question of whether a Bayesian/utilitarian or maximin/contractarian analysis is better for all situations. My own opinion is (1) that Bayesian rules are better in many cases of $risk_B$ or certainty and in many situations of *individual* (not societal) risk-taking, and (2) that both Bayesian and maximin strategies are needed in moral philosophy. However, I shall not take the time here to argue these points in detail. As Amartya Sen points out, utilitarianism and contractarianism capture two different aspects of welfare considerations. Both provide necessary conditions for ethical judgments, but neither alone is sufficient. Utilitarians are unable to account for how we evaluate different *levels* of welfare, and contractarians are unable to account for how to evaluate gains and losses of welfare.[17]

Insofar as the controversy between utilitarians and contractarians or egalitarians in moral philosophy is relevant to hazard assessment and to democratic decisionmaking, the main question at issue is how a rational person ought to behave in situations of *societal* (not individual) risk under uncertainty. A reasonable way to answer this question is to examine carefully the best contemporary defenses, respectively, of the Bayesian and maximin positions, and then apply these examinations to various cases of technological/environmental risk, in which society must make a decision under uncertainty. The best of these defenses are probably provided by Harsanyi, an act utilitarian, and Rawls, a contractarian and egalitarian.

3. HARSANYI'S ARGUMENTS

Harsanyi's main arguments in favor of the Bayesian and against the maximin strategy under uncertainty are as follows: (1) Those who do not follow the Bayesian strategy make irrational decisions because they ignore probabilities. (2) Failure to follow the Bayesian strategy leads to irrational and impractical consequences. (3) This failure also leads to unacceptable moral consequences. (4) Using the Bayesian strategy and Harsanyi's equiprobability assumption is desirable because it allows one to assign equal *a priori* probability to everyone's interests. (5) If one fails to use the Bayesian strategy, one irrationally presupposes that the worst outcome has a probability of one.

3.1 The First Argument for the Bayesian Strategy:
Since an adequate consideration of the Rawls-Harsanyi debate requires an analysis of the arguments and counterarguments made by each of them, a task too extensive for this short essay, let us focus on Harsanyi's first argument. This is enough to show the flaws in his position. The argument is that choosing the maximin stategy would lead to irrational decisions and to total risk aversion. Such aversion is irrational, he says, since it always requires one to avoid the worst possible consequences, regardless of their probability of occurrence. The main assumption underlying this argument is, in Harsanyi's words: "It is extremely irrational to make your behavior wholly dependent on some highly unlikely unfavorable contingencies, regardless of how little probability you are willing to assign to them."[18]

To substantiate his argument that ignoring probabilities would lead to total risk aversion and hence to an irrational position, Harsanyi gives an example of maximin decisionmaking and alleges that it leads to paradoxes. The example is this. Suppose you live in New York City and are offered two jobs, in different cities, at the same time. The New York City job is tedious and badly paid, and the Chicago job is interesting and well paid. However, to take the Chicago job, which begins immediately, you have to take a plane, and the plane travel has a small, positive, associated probability of fatality. This means, says Harsanyi, that following the maximin principle would cause you to accept the New York job, as represented by the following table:

	If the Chicago plane crashes	If the Chicago plane does not crash
If you choose New York job	Then you have a poor job but will be alive.	Then you have a poor job but will be alive.
If you choose Chicago job	Then you will die.	Then you have a good job but will not die.

In the example, Harsanyi assumes that your chances of dying in the near future from reasons other than a plane crash are zero. Hence he concludes that, because maximin directs choosing so as to avoid the worst possibility, therefore it forces one to ignore both the low probability of the plane crash and the desirability of the Chicago job and instead to choose the New York job. However, Harsanyi claims that a rational person, using the expected-utility criterion, would choose the Chicago job for the very two reasons, viz., its desirability and the low probability of a plane crash on the way to Chicago.

3.2 Problems with the Argument:

How successful is Harsanyi's argument in employing the counterexample of the New York and Chicago jobs? There are a number of reasons why it is inadequate to establish the thesis that in facing technological risk under uncertainty, one ought to use Bayesian rather than maximin decision strategies. For one thing, the example is highly counterintuitive; it is hard to believe that the greatest risk comes from dying in a plane crash, since hazard assessors have repeatedly confirmed that the average annual probability of fatality associated with many other activities, e.g., driving an automobile, is greater by an order of magnitude than that associated with airplane accidents.[19] Hence Harsanyi has stipulated, contrary to fact, that the worst case would be dying in a plane crash. This means that his stipulation, not use of the maximin rule, could be the source of much of what is paradoxical about the example.

Second, even if the example in this argument were plausible, it would prove nothing about the undesirability of using maximin in situations of *societal* risk under uncertainty, like deciding whether to build a nuclear reactor or to open a liquefied natural gas facility. Harsanyi makes the assumption, in using this example, that the situation of uncertainty regarding *one* individual's death, caused by the same person's decision to fly to Chicago, is no different than a situation of uncertainty regarding *many* individuals' deaths, caused by a societal decision to employ a hazardous technology, e.g., a nuclear reactor. In the *individual* case, the risk is typically freely chosen, but in the *societal* instance, it is often involuntarily imposed. Despite this assymmetry in the two cases, Harsanyi assumes that rational decisions are defined in the same way for both instances.

3.21 The Individual and Societal Cases Are Disanalogous:

Objecting to Harsanyi's example, Rawls made a point similar to mine; he claimed that the example failed because it was of a small-scale, rather than a large-scale, situation.[20] My claim is similar, but more specific: situations of individual risk are voluntarily chosen, whereas situations of societal risk are typically involuntarily imposed; hence they are not analogous. This means that, to convince us that *societal* decisions in situations of uncertainty are best made by following a Bayesian rule, Harsanyi cannot merely provide an example of an *individual* decision. Otherwise he ignores the democratic constraints on societal risk imposition.

Harsanyi, however, disagrees. Answering Rawls's objection, he claims that, "Though my counterexamples do refer to small-scale situations, it is very easy to adapt them to large-scale situations since they have instrinsically nothing to do with scale, whether large or small. . . . I cannot see how anybody can propose the strange doctrine that scale is a fundamental variable in moral philosophy."[21]

There are several reasons why Harsanyi is wrong in this claim. In the *individual* case, one has the right to use expected utility so as to make efficient, economical decisions regarding *oneself*. In the *societal* case, one does not always have the right to use expected utility so as to make efficient, economical decisions regarding *others* in society, since maximizing utility or even averaging utility might violate rights, duties, or obligations. Instead, one may have democratic duties to members of society; for example, even when they do not recognize the existence of the duties, and even when they are ignorant of their own best interests.

On the *individual* level, the question is *why* the individual can define risk in a certain way, if he is to be said to be rational, and if his concept of rationality is to be theoretically justifiable. On the *societal* level, the question is *how* a group of people can define risk in a certain way, if they are to be said to be culturally rational, and if their concept of cultural rationality is to be *democratically* justifiable, in terms of ethical procedure.

If ethical procedure is essential to societal welfare judgments, then the Bayesian strategy may be irrational at the most fundamental level. This is because if rational societal decisionmaking requires an ethical rule that takes account of the fairness of the allocational *process*, not merely the *outcomes*, then the Bayesian strategy cannot provide such a rule.[22] It cannot do so because of what Sen calls "the strong independence axiom,"[23] viz., what Harsanyi calls the "sure thing principle" or the "dominance principle." This axiom essentially states that, if one strategy yields a better *outcome* than another does under some conditions, and it never yields a worse outcome under any conditions, then decisionmakers always ought to choose the first strategy over the second.[24] If there are grounds for doubting the sure-thing principle, because it ignores *ethics and democratic process* and focuses only on outcomes, then there are equally important grounds for doubting Bayesianism. This is because the sure-thing principle is one of the three main rationality axioms underlying the Bayesian approach. As Harsanyi puts it, "The Bayesian approach stands or falls with the validity or invalidity of its *rationality axioms.*"[25]

If the sure-thing principle fails, especially in the societal case, to provide a full account of rational behavior, then there are even stronger grounds for questioning Bayesianism in situations of decisionmaking under uncertainty. This is because democratic process is probably more important in cases where probabilities are unknown than in those where they are certain, since it would be more difficult to insure informed consent in the former cases. This, in turn, suggests that if Bayesianism has trouble with defining rational moral behavior in the societal case, then it may need a modified definition of rationality. The modification is needed to address the fact that the individual case has to do more with a *substantive* concept of rationality, whereas the societal case has more to do with a *procedural* or "process" concept of rationality,[26] since it must take account of conflicting points of view, various ethical and legal obligations, and so on. In other words, it must take account of democracy.

For example, I may have an obligation to help insure that all persons receive equal protection under the law, even if a majority of persons are unaware of their rights to equal protection, and even if their personal utility functions take no account of these unknown rights. In other words, if I make a decision regarding my own risk, I can ask "How safe is rational enough?" and I can be termed irrational if I have a fear of flying. But if I make a decision regarding risks to others in society, I do not have the right to ask, where their interests are concerned, "How safe is rational enough?" Or in the individual case, we might ask, "How safe is safe enough?" as a norm for rationality. In the societal case, however, we must ask, because we are bound by moral obligation to others, "How safe is free enough?" or "How safe is fair enough?" or "How safe is voluntary enough?"

But if people attempting to make rational societal decisions are bound by ethical obligations, then this suggests that there might be what W. K. Clifford and Alex Michalos have called "the ethics of belief."[27] I shall not take the time to outline arguments for the negative and positive ethics of belief in this essay. But it is important to point out that, if these arguments are correct, then they provide additional grounds for believing that rationality, especially societal rationality, has a moral and democratic dimension. Even Harsanyi recognizes this, in some sense, since he divides the general theory of rational behavior into three branches: utility theory, game theory, and ethics. The first branch deals with individual rational behavior under certainty, risk, and uncertainty; it defines "rational behavior" in terms of expected utility maximization. Game theory deals with the rational behavior of two or more individuals, each with utility functions. Ethics, for Harsanyi, is the theory of rational moral judgments, where rational moral judgments are defined in terms of maximizing the average utility level of all individuals in society.[28] Moreover, although Harsanyi explicitly claims that his utility theory is applicable to ethics, it can easily be argued that efficiency and utility alone may not be capable of dealing with the most important dimensions of ethics.

3.22 Harsanyi Begs the Question:

A third problem with Harsanyi's argument is that he begs the very question he sets out to prove, viz., that maximin leads to irrational consequences in situations of risk under uncertainty. Harsanyi writes: "It is extremely irrational to make your behavior [taking the job in New York] wholly dependent on some highly unlikely unfavorable contingencies [the Chicago plane crash], regardless of how little probability you are willing to assign to them."[29] On Harsanyi's own admission, he is attempting to prove that maximin ought not be used in situations of risk under *uncertainty*.[30] Yet, if the worst consequences, death in a plane crash, are *by his own definition*, "highly unlikely,"[31] then these consequences have a *stipulated* low probability. Indeed, we know the likelihood of plane crashes for different aircrafts. The situation is not one of uncertainty, but one of $risk_B$. This means that, at best, Harsanyi's example shows that it is not reasonable, in a situation of $risk_B$, to forego great benefits (a good job in Chicago) in order to avoid a very small probability of death (a plane crash), particularly if other causes of death, at other jobs, are more likely. Harsanyi's example has proved nothing about rational decisions under *uncertainty*.

A related problem with Harsanyi's account is that he claims that people behave *as if* they maximized utility. In situations of uncertainty, this claim begs the question because it cannot be demonstrated. It cannot be demonstrated because we do not know the relevant probabilities.

3.23 Problems with Linearity and Gambling:

Even as a strategy for rational decisions under $risk_B$, however, it is not clear that Harsanyi's principles would always be successful. He claims that reasonable people do not forego great benefits in order to avoid a small probability of harm. This statement presupposes that people want to gamble, and that there are reasonable ways of gambling. However, it appears equally plausible to argue that many rational people do not wish to gamble, especially if their lives are at stake. Why should one gamble *with one's life* unless it is an absolute necessity? For example, why should one choose to avoid an airplane delay (a benefit), at the risk or cost of facing a 10^{-6} probability that an essential mechanism on the plane will break down? A perfectly rational response, in such a situation, might be that one does not gamble with one's life except to obtain a comparably great benefit.[32]

Moreover, in the cases in which we might justifiably gamble (with something other than our life), it is not clear that humans are Bayesians at all. J. W. N. Watkins gives an interesting example to illustrate this point:

Consider a certain result r and a family of gambles G_p, defined as getting r with probability p and nothing otherwise. Classical utility theory evaluates the utility of G_p as p.u(r). Since u(r) is constant this means that $u(G_p)$ is a linear function of p. In particular, equal increases in p mean equal increases in $u(G_p)$. But do we not evaluate the step from $G_{0.99}$ to $G_{1.00}$ higher than e.g. from $G_{0.5}$ to $G_{0.51}$?

Watkins's example shows that for the same result, e.g., avoiding a nuclear core melt, the result is not a linear function of p, with the result becoming more desirable in direct proportion as the probability of avoidance increases. Instead, says Watkins, the jump from a probability of 0.99 to 1.00 is worth more than the jump from, say, 0.50 to 0.51, since the first jump represents the value attached to complete avoidance of the risk. But if so, then rational decisionmakers are not necessarily Bayesian. Their risk aversion is not necessarily a linear function of probability.[33] And if not, then we have grounds for doubting the Bayesian strategy.

3.24 Problems with Subjective Probabilities:

Likewise, Daniel Kahneman and Amos Tversky, for example, have argued persuasively that the Bayesian model does not capture the essential determinants of the judgment process. This is because they found, for instance, that sample size has no effect on subjective sampling distributions, that posterior binomial estimates are determined by sample proportion rather than by sample difference, and that they do not depend on the population proportion. Kahneman and Tversky showed that the same types of systematic errors, suggested by considerations of representativeness, "can be found in the intuitive judgments of sophisticated scientists."[34] This suggests that, if subjective probabilities are frequently prone to error, then, contrary to Harsanyi's argument, rational people could well decide to avoid them, at least in certain situations.

3.25 Why Not Base Some Decisions on Consequences?

Harsanyi's argument is problematic, sixth, because it is built on the supposition that "It is extremely irrational to make your behavior wholly dependent on some highly unlikely unfavorable contingencies regardless of how little probability you are willing to assign to them."[35] What Harsanyi is saying is that it is irrational to base decisions on consequences and to ignore the probability associated with them. However, it is not clear that it is irrational to avoid highly unfavorable contingencies, even if they are associated with only a small probability.

Suppose one has the choice between buying organically grown vegetables and those treated with pesticides. Suppose, second, that the price difference between the two is very small. And suppose, third, that the probability of getting cancer from the vegetables treated with pesticide is "highly unlikely," to use Harsanyi's own words. It is not irrational to avoid this cancer risk, even if it is small, particularly if one can do so at no great cost. Even more obvious is that it is not irrational to avoid a catastrophic risk, e.g., nuclear winter, even if it were small. Some consequences are so undesirable, so catastrophic, that it makes sense to try to avoid them, regardless of their probability, and especially if one can do so at no great cost.

Moreover, were one to accept Harsanyi's assumption, that she ought to ignore "highly unlikely" contingencies, then it is reasonable to believe that various industries and manufacturers would succeed in imposing a number of small risks on the public, risks that together might amount to a substantial hazard.

4. CONCLUSION

The upshot of my criticisms of Harsanyi's argument is that it has failed on at least six counts. The most important is that Harsanyi's use of an *individual* decision (to risk one's own life) as an analogue for a *societal* decision (to risk the lives of many people, most of whom have no real voice in the choice) is highly questionable. It ignores some basic democratic constraints on public decisionmaking, namely, equity of risk distribution; free, informed consent to risk; and due process. For Harsanyi to have a real counterexample to using maximin in situations of societal risk under uncertainty, he would have to come up with a case in which someone caused disastrous or questionable consequences precisely because she used maximin rules in societal decisionmaking under uncertainty. This he has not done. Moreover, if the earlier arguments are correct, Harsanyi cannot provide such an example. He cannot do so because he has no *procedural* account of rationality. Without such an account, he misses part of what is essential to democracy.

University of South Florida

NOTES

1. U.S. Nuclear Regulatory Commission, *Reactor Safety Study*, NUREG-75/014, WASH 1400 (Washington, D.C.: U.S. Government Printing Office, 1975); K. Shrader-Frechette, *Nuclear Power and Public Policy* (Dordrecht: Reidel, 1983), pp. 84-85.
2. Union of Concerned Scientists, *The Risks of Nuclear Power Reactors, a Review*

of the NRC Reactor Safety Study WASH 1400 (Cambridge, Mass.: Union of Concerned Scientists, 1977); Nuclear Energy Policy Study Group, *Nuclear Power: Issues and Choices* (Cambridge, Mass.: Ballinger, 1977); this report was funded by the Ford Foundation and carried out by the Mitre Corporation; R. M. Cooke, "Risk Assessment and Rational Decision," *Dialectica* 36:4 (1982): 334.

3. See note 2.

4. See Cooke, "Risk Assessment."

5. For this and other decision rules, such as minimax regret, see M. Resnick, *Choices* (Minneapolis: University of Minnesota Press, 1987), pp. 26-37.

6. J. Harsanyi, "Can the Maximin Principle Serve as a Basis for Morality? A Critique of John Rawls's Theory," *American Political Science Review* 59:2 (1975): 594; hereafter cited as: Harsanyi, 1975.

7. J. Harsanyi, "Advances in Understanding Rational Behavior," in J. Elster, ed. *Rational Choice* (New York: New York University Press, 1986), p. 88; hereafter cited as: Harsanyi, 1986; J. Harsanyi, "Understanding Rational Behavior," in R. Butts and J. Hintikka, eds., *Foundational Problems in the Special Sciences* (Dordrecht: Reidel, 1977), vol. 2, p. 322; hereafter cited as: Harsanyi, 1977. See also A. Tversky and D. Kahneman, "The Framing of Decisions and the Psychology of Choice," in Elster, *Rational Choice,* p. 125.

8. See Resnick, *Choices,* p. 32.

9. Harsanyi, 1975, p. 594; Harsanyi, 1977, pp. 320-321. For a treatment of individual decisions under uncertainty, see R. Luce and H. Raiffa, *Games and Decisions* (New York: Wiley, 1957), pp. 275-326.

10. H. Otway and M. Peltu, *Regulating Industrial Risks* (London: Butterworths, 1985), p. 4.

11. Harsanyi, 1977, p. 320; Otway and Peltu, *Regulating Industrial Risks,* p. 115; Resnick, *Choices,* p. 36; see pp. 88-91 for the Von Neumann-Morgenstern approach to expected utility.

12. Harsanyi, 1975, p. 594; see also R. C. Jeffrey, *The Logic of Decision* (Chicago: University of Chicago Press, 1983); J. Marschak, "Towards an Economic Theory of Organization and Information," in R. Thrall, C. Coombs, and R. David, eds. *Decision Processes* (London: Wiley, 1954), pp. 187 ff.; and L. Ellsworth, "Decision-Theoretic Analysis of Rawls' Original Position," in C. Hooker, J. Leach, E. McClennen, eds. *Foundations and Applications of Decision Theory* (Dordrecht: Reidel, 1978), pp. 39 ff. Two of the scholars who are most vocal in the claims that utility is the way to analyze choice situations involving uncertainty are Robert Davis, "Introduction," in Thrall, Coombs, and Davis, pp. 14, and C. Coombs and D. Beardslee, "On Decision-Making Under Uncertainty," in the same volume, pp. 255 ff. For an opposed point of view, see E. F. McClennen, "The Minimax Theory and Expected Utility Reasoning," in Hooker, Leach, and McClennen, pp. 337 ff. McClennen argues that there is a strong reason to doubt that the expected utility principle is a basic criterion for rational behavior. For a review of a number of decisionmaking procedures, see J. Milnor, "Games against Nature," in Thrall, Coombs, and Davis, pp. 49 ff.

13. Harsanyi, 1977, p. 323; for more references to the work of Harsanyi on this point see J. Harsanyi, "On the Rationale of the Bayesian Approach," in Butts and Hintikka, *Foundational Problems;* hereafter cited as: Harsanyi, 1977a. See Resnick, *Choices,* p. 36; for a definition, discussion, and proof of the expected-utility theorem of Von Neumann-Morgenstern, see Resnick, pp. 88 ff. See Harsanyi, 1977, pp. 320-322, for a brief discussion of Bayesian decision theory and its associated axioms.

14. Harsanyi, 1975, p. 595; for a discussion of the maximin rule, see Resnick, *Choices,* pp. 26 ff.; for a discussion of individual decisionmaking under uncertainty, including maximin, see Luce and Raiffa, *Games and Decisions,* chapter 13.

15. J. Rawls, *A Theory of Justice* (Cambridge, Mass.: Harvard University Press,

1971), pp. 75-83.
16. This example is based on Resnick, *Choices*, p. 41. For a treatment of group decisionmaking, see Luce and Raiffa, *Games and Decisions*, pp. 275- 326.
17. A. K. Sen, "Rawls Versus Bentham: An Axiomatic Examination of the Pure Distribution Problem," in N. Daniels, ed., *Reading Rawls* (New York: Basic Books, 1981), pp. 283-292.
18. Harsanyi, 1975, p. 595.
19. See C. Starr and C. Whipple, "Risks of Risk Decisions," *Science* 208, no. 4448 (1980): 1118. See also J. Hushon, "Plenary Session Report," in *Symposium/Workshop on Nuclear and Nonnuclear Energy Systems: Risk Assessment and Governmental Decision Making* (McLean, Va.: Mitre Corporation, 1979), p. 748. See also D. Okrent, "Comment on Societal Risk," *Science* 208, no. 4442 (1980): 374. See M. Maxey, "Managing Low-Level Radioactive Wastes," in J. Watson, ed., *Low-Level Radioactive Waste Mangement* (Williamsburg, Va.: Health Physics Society, 1979), p. 401. See J. D. Graham, "Some Explanations for Disparities in Lifesaving Investments," *Policy Studies Review* 1:4 (May 1982): 692-704, as well as C. Comar, "Risk: A Pragmatic De Minimus Approach," *Science* 203, no. 4378 (1979): 319. Finally see B. Cohen and I. Lee, "A Catalog of Risks," *Health Physics* 36:6 (1979): 707.
20. J. Rawls, "Some Reasons for the Maximin Criterion," *American Economic Review* 64 (May 1974): 141-146, esp. p. 142.
21. Harsanyi, 1975, p. 605.
22. P. Diamond, "Cardinal Welfare, Individualistic Ethics, and Interpersonal Comparisons of Utility," *Journal of Political Economy* 75 (1967): 765-766.
23. A. Sen, "Welfare Inequalities and Rawlsian Axiomatics," in Butts and Hintikka, *Foundational Problems*, p. 276.
24. Harsanyi, 1977a, p. 384, and Harsanyi, 1977, p. 321.
25. Harsanyi, 1977a, p. 382.
26. See J. G. March, "Bounded Rationality," in Elster, *Rational Choice*, p. 148, for a discussion of "process rationality."
27. W. K. Clifford, *Lectures and Essays* (London: Macmillan, 1986); A. C. Michalos, *Foundations of Decisionmaking* (Ottowa: Canadian Library of Philosophy, 1987), pp. 204-218.
28. Harsanyi, 1975; Harsanyi, 1977, p. 323; Harsanyi, 1977a; Harsanyi, 1986.
29. Harsanyi, 1975, p. 595.
30. Harsanyi, 1975, p. 594.
31. Harsanyi, 1975, p. 595.
32. See J. W. N. Watkins, "Towards a Unified Decision Theory: A Non-Bayesian Approach," in Butts and Hintikka, *Foundational Problems*, p. 351.
33. *Ibid.*, p. 375.
34. D. Kahneman and A. Tversky, "Subjective Probability," in D. Kahneman, A. Tversky, and P. Slovic, eds., *Judgment under Uncertainty: Heuristics and Biases* (Cambridge: Cambridge University Press, 1981), p. 46; see also D. Kahneman and A. Tversky, "On the Psychology of Prediction," in the same volume, p. 68, where they show that even statistical training "does not change fundamental intuitions about uncertainty."
35. Harsanyi, 1975, p. 595.

RICHARD E. SCLOVE

THE NUTS AND BOLTS OF DEMOCRACY: DEMOCRATIC THEORY AND TECHNOLOGICAL DESIGN

Technology is a taboo subject in philosophy and the social sciences. Yes, there is an occasional passing mention. Even a seminal book or two. But the vast mainstream of social science proceeds as though technologies simply do not exist, do not matter.

The principal exception occurs within the discipline of economics, in which a modest scholarly industry now preaches the need to devise public policies that can accelerate the pace of technological innovation.[1] The stated objective is to enhance national economic growth, productivity, and international competitiveness—on the unexamined assumption that as long as an innovation sells profitably, it is an unalloyed social blessing.

Astonishing. It is as perplexing as if for generations a family shared its home with a giant, orange, seizure-prone elephant, and yet never discussed—somehow never even noticed—the beast's presence and pervasive influence upon every facet of their lives. It is even as though everyone in the United States gathered together each night in their dreams—assembled solemnly in a glistening moonlit glade—and there furiously debated, radically rewrote, and ratified a new Constitution. Then awakening with no memory of what had passed, they yet spent each next day mysteriously complying with the nocturnally revolutionized document in its every word and letter.

Such a world, such a wildly impossible world in which unconscious collective actions govern waking reality, is the world that now exists.[2] It is the modern technological world that we have all—albeit some more potently or self-seekingly than others—helped create. Hence my central message: We must learn to take this essay's title not only seriously, but literally. The "nuts and bolts of democracy"—ordinarily a metaphor through which we denote concern with the practical arrangement and governance of a democratic society—must grow to encompass a literal concern with nuts and bolts. An engaged citizenry must become critically involved with the choice, governance, and even design of technological artifacts and practices. Citizens must furthermore grow committed to adopting only technologies that are compatible with preserving or strengthening their society's democratic nature. Or else—like hapless daytime victims of our own midnight cabals—we can have no democracy, or none worthy of the name.

I have elsewhere outlined practical steps for achieving a democratic politics of technology.[3] Here I wish to backtrack and explore reasons for undertaking such a program.

TECHNOLOGIES AS SOCIAL STRUCTURES

First, we must recognize that technologies represent an important species of social structure, functionally comparable to other sorts of social structures such as major economic institutions, laws, or systems of cultural belief. Take an example.

During the early 1970s running water was installed in the houses of Ibieca, a small village in northeast Spain. With pipes running directly to their homes, Ibiecans no longer had to fetch water from the village fountain. As families gradually purchased washing machines, fewer village women gathered to scrub laundry by hand at the village wash basin.

Arduous tasks were thus rendered technologically superfluous. But village social life was also transformed, and in unintended ways. The public fountain and wash basin, once sites of bustling productive and social interaction, became nearly deserted. Men lost some of their sense of familiarity with children and donkeys that had helped them haul water. Women stopped gathering at the wash basin to intermix scrubbing with politically empowering gossip about the men and village life generally.

In hindsight this emerges as a vital step in a broader historical process through which Ibiecans came to lose the strong experiential and emotional bonds—with one another, animals, and the land—that had formerly knit them together into a community.[4] In the United States the automobile has played a somewhat similar role in disrupting earlier patterns of community self-identification and bonding.[5]

We can use the simple example of Ibieca to illustrate some important generalizations concerning the role of technologies as social structures. First, technologies indeed help *structure social relationships*. In the case at hand, upsetting a traditional pattern of water use compromised important means through which a village had formerly reproduced itself through time as a self-conscious community. Second, technologies tend to do this *independently of their (nominally) intended purposes*. Fountains, pipes (or, for that matter, microwave ovens, hypodermic syringes, garden hoses, or numerically-controlled machine tools) are not normally regarded as devices designed to shape patterns of human relationship, but that is nevertheless one of their pervasive tendencies. Third, in saying that technologies "structure" social relationships, I mean that they *shape, but do not fully determine*, the nature of social reality.[6] Water pipes and washing machines did not, for example, materially force Ibiecans to stop gathering at the central fountain, but instead altered the system of inducements and interdependencies that had formerly made such gathering a seemingly natural occurrence. Fourth, technologies are themselves *contingent social products*.[7] No grand law of evolution dictated that pipes and washing machines would be invented and marketed in the form they were, or that Ibieca would adopt the new technologies that it did, or when it did. One obvious alternative would have been to build a community laundry facility. All such innovations are invariably a result of explicit or tacit processes of social choice, conditioned by a prevailing complex of social structures and forces, including the pre-existing technological order.

Along with their material influence upon social experience, technologies also exert symbolic and other cultural influences. For instance, nineteenth-century New England factories and schools architecturally resembled churches, complete with rectangular bell towers. One social result, whether or not this was the designers' conscious intent, was to help symbolically impress upon workers and children the need to maintain at all times the same quiet, order, and deference to authority that they were taught to exhibit at worship.[8]

Technologies even play transformative roles within psychological development. For example, earlier in this century Swiss psychologist Jean Piaget determined that young children distinguish living from non-living things according to whether or not the things move, and—as the children continue to develop psychologically—then according to whether things move by themselves or are moved by an outside force or agent. But more recently social psychologist Sherry Turkle has discovered that children who play with computer toys that appear to "talk" and "think," develop different criteria for distinguishing alive from not alive. Instead of relying on physical criteria (e.g., motion), they invent psychological criteria and hypotheses (e.g., "Computers are alive because they cheat" or "Computers are not alive because they don't have feelings"). Children's developmental trajectories, including their self-conceptions and moral reasoning, are thus transformed as a result of their interactions with these artifacts.[9]

The process that Turkle describes with respect to computer toys is, moreover, only a specific instance of a much more general phenomenon.[10] In reconfiguring our opportunities and constraints for action, and in functioning simultaneously as symbols and latent communicative media, technologies also reconfigure our opportunities and constraints for psychological development, including self-actualization.

In short, in contrast with the prevailing view that technologies are simply "tools" or "instruments" that accomplish a single pre-determined or "focal" end, I suggest that in addition—and often more importantly—technologies help both focally and non-focally to shape patterns of social organization and interaction, cultural evolution, psychological development, and generally constitute much of the world in which we live. Such insights lead me to regard technologies as an important class of social structure.

I use the term "technology" broadly to encompass roughly what anthropologists call "material culture." That is, materially-embodied human artifacts (including architecture), along with the social practices that accompany—or are influenced by—the production or existence of these artifacts.

A general property of technologies is that, beyond their designed intent, each typically performs an indefinitely large number of further functions, helps induce an indefinitely large number of effects, and conveys or embodies an indefinitely large number of meanings and symbolic values. I denote this tendency by saying that technologies are "polyvalent" in their functions, effects, and meanings.

If we designate a technology's (nominally) intended or designed purpose as its "focal function," then we can refer to its accompanying complex of further—but often experientially recessive—effects and attributes as its *non-focal* functions, effects and meanings. Thus the focal purpose of nineteenth-century New England schoolhouses was to provide a space for educational instruction, while one of their non-focal functions was to help generate—in part via architectural symbolism—a relatively docile and obedient workforce.

It is generally in virtue of their polyvalency—including, often most importantly, their non-focal aspects—that technologies come to function as social structures. For instance, in the case of Ibieca the focal function of water pipes was to distribute water directly to homes; one accompanying non-focal effect was to render the village water basin and fountain materially superfluous in daily life—a result that in turn helped produce a concatenation of further culturally significant results. And it is in virtue especially of these non-focal effects that the pipes qualify as a social structure.

Technologies are not the only social structures. Others include dominant political and economic institutions, laws, cultural belief systems and language. Each of these—like technologies—qualify as social structures in virtue of being contingent social products. Then, once they come into existence, they tend to endure and profoundly influence (without, however, fully determining) one another's subsequent evolution, as well as the course of history and texture of daily experience.

If, compared with other social structures, technologies warrant distinctive concern, it is owing primarily to our relative obliviousness to the twin fact of their contingent social origins and subsequent structural significance. This acquired blindness enhances technologies' relative structural importance, for it enables them to exert their influence with only limited social awareness of how—or even that—they are doing so.

TECHNOLOGY AND DEMOCRACY

But what has this got to do with democracy, and what is so important about democracy anyway? Simply put—and here I rely on a reconstruction of the political philosophies of Rousseau and Kant—democracy is a necessary background condition for enabling people to develop individual moral autonomy, and to debate and decide together whatever else, aside from democracy, should matter to them.[11] Democracy is in this sense a highest-order shared human value, because it is fundamental—crucial to formulating and realizing other values.

This suggests a moral obligation to help establish democratic social orders, whose institutional essence should be to provide members with relatively equal and extensive opportunities to determine the collective conditions of their existence. However, an important *a priori* constraint upon such political involvement is that citizens should reproduce through time the democratic nature of their society.

Implications for technological design and practice emerge straightforwardly. If the essence of democracy is that citizens ought to be empowered to participate in determining the collective conditions of their existence, particularly as these bear upon the possibility of democracy itself, and if technologies are themselves social structures, it follows that it is morally imperative to democratize technological design and practice.

Meaning what? Technological democratization would have both procedural and substantive components. Procedurally, there must be expanded opportunities for people from all walks of life to participate effectively in guiding the evolving technological order. Substantively, the resulting technologies ought to be compatible in their design with democracy's necessary conditions.

This is a simple argument. With respect to contemporary technological trends, its implications are also thoroughly radical. Assuming its validity, it is morally imperative—necessary to our own freedom and dignity—to take processes of technological development that are today guided primarily by market forces, distant bureaucracies, economic self-interest, or international rivalry, and instead subordinate them to democratic prerogatives.

I am not insisting that technologies are necessarily the most important factor that influences political life. Rather, they are sufficiently important—and so inextricably intertwined with other such factors—that we must learn to subject them to the same sort of rigorous political scrutiny that we understand should be accorded to those others (such as laws, international relations, the distribution of wealth, and so on).

The notion that there ought to be sovereign popular participation in evaluating and choosing technologies is perhaps familiar.[12] Less familiar is the idea that, in addition, there must be extensive popular participation in the prior process of technological design—and that designs themselves must be substantively compatible with democracy. For this reason, I wish to focus especially on the implications of democratic theory for technological design.

DESIGN CRITERIA FOR DEMOCRATIC TECHNOLOGIES

Let us consider first the necessity of seeking technological designs that are substantively compatible with democracy's necessary conditions. What are those conditions?

In concrete terms, they will vary somewhat depending upon historical circumstances and cultural preferences, but let us suppose that three general institutional requirements must be: (1) democratic politics, (2) democratic communities, and (3) democratic work. For a schematic definition of these concepts and their practical interrelationships, see figure 1:[13]

Several examples can help us see ways in which technologies can affect the establishment of such democratically necessary conditions.

Technology and Democratic Politics:

Certain industrial facilities—such as nuclear power plants and hydroelectric dams—have been the target of terrorist threats and, in some instances, actual assaults. In 1980, for instance, armed men, besieging the embassy of the Dominican Republic in Bogota, Columbia, attempted to ward off U.S. intervention by issuing a communiqué that read: "You must remember, U.S. gentlemen, that you have never experienced war in your vitals, and that you have many nuclear reactors."[14]

Catastrophic sabotage of an operating nuclear power plant could cause a disaster of Chernobyl-like (or worse) proportions, and prompt

Figure 1. *Necessary Conditions of Democracy*

"Democratic politics," "democratic community," and "democratic work" are three organizational principles that ought, with varying individual intensities, to be present in every basic institutional setting within a democratic society:

DEMOCRATIC POLITICS

a. Complementary participatory and representative institutions, within a context of egalitarian, globally-aware political decentralization and federation. (Representative institutions designed to support and incorporate direct citizen participation.)
b. Respect for essential civil rights and liberties.

To help establish equal respect, collective efficacy, and commonalities:

DEMOCRATIC COMMUNITY

a. Face-to-face human interaction on terms of equality as a means to nurture mutual respect, emotional bonds, and recognition of commonalities among citizens.
b. Intercommunity cultural pluralism.
c. Extensive opportunities for each citizen to hold multiple memberships across a diverse spectrum of communities.

To help develop citizens' moral autonomy, including their capacities to participate effectively in politics, and the propensity to grant precedence to important common concerns and interests:

DEMOCRATIC WORK

a. Equal and extensive opportunities to participate in self-actualizing work experiences.
b. Diversified careers, flexible life-scheduling, and citizen sabbaticals.

subsequent reduction or suspension of essential civil liberties and rights, as governments sought to ferret out subversive groups capable of engineering other such calamities.[15] Thus certain technologies may pose a standing threat to democratic politics. At the same time, alternative energy technologies are available which are not susceptible to catastrophic sabotage,[16] and which also lack the unique symbolic overtones that are culturally invested in nuclear power, demonstrably heightening its attractiveness to potential saboteurs.

Technology and Democratic Community:
Many factors have played a role in the modern decline of face-to-face community, but important among them have been technological developments such as the private automobile, detached single-family homes, television, and the loss of public space (e.g., town commons supplanted by shopping malls).[17] In Zurich, Switzerland, a partial antidote is being sought through a government program that provides neighborhoods with special community-focused encouragement, legal advice, and architectural counseling. As a result neighbors have begun to collaborate in removing backyard fences; building new walkways, gardens and other neighborhood facilities; and thus generally refashioning private yards into a well-balanced blend of private, semi-public, and public spaces.[18]

Technology and Democratic Work:
Some years ago sociologist (and one-time union organizer) Robert Schrank discussed alternative work arrangements with a group of General Motors Corporation union representatives. After describing Scandinavian experiments in redesigning factories to permit more interesting work and greater worker involvement in governing the day-to-day labor process, Schrank asked the men to imagine how they would redesign their own factories if they had a chance. Their response was skeptical and unenthusiastic. Later Schrank reflected: "The frame of reference of these workers was the linear assembly line as they experienced it. Even to think beyond that seemed difficult."[19]

Thus linear assembly lines not only tend to restrict possibilities for worker self-management, conviviality, and meaningful work. They can also exert an ideological force, impairing workers' abilities even to envision technological alternatives. And yet, as Schrank was eventually able to persuade these particular workers, more democratic automobile manufacturing technologies have been in use for some years. For example, an innovative Volvo factory in Kalmar, Sweden, uses independently-movable electronic dollies—each carrying an individual auto chassis—in place of a traditional assembly line. The dollies enable small teams of workers to help plan and vary their daily work routines, while assembling entire automobile subsystems.

By using democratic theory to interpret and abstract from such case studies, I have generated a provisional list of criteria to help distinguish relatively democratic technologies from those that are non-democratic.[20] I characterize these criteria as rationally contestable, because I do not consider this a complete or definitive list. Rather, I envision that by debating the formulation and use of such criteria, citizens could gradually evolve democratic *processes* of decisionmaking, that would issue in technologies that were themselves *substantively* democratic in their design and use. See figure 2.

Each of these eleven criteria is intended to have some direct and important bearing upon democracy's possibility. In making technological decisions, there are of course many other issues that we might want to address. But my belief is that we should attend first and especially to democracy. Not because it is the only thing that matters, but because it provides the necessary background for being able to decide freely and fairly what other considerations to take into account. Within such a process, democratic deliberative forums must retain both formal and substantive sovereignty over expert testimony, because only the former can supply both the intersubjectively balanced impartiality and full range of social knowledge that legitimate determinations require.

COMPARISON WITH ECONOMIC THEORY

To better grasp its significance, we may briefly contrast the idea of using democratic design criteria to help choose among alternative technologies with one of its principal competitors: neo-classical economics, as represented by such methodologies as applied welfare economics, cost-risk-benefit analysis, and decision analysis.

Neo-classical welfare economics screens technologies and offers policy prescriptions based on an assumption of one-directional causality: the basic structure of any society—that is, the political and cultural background conditions that (among other things) regulate markets and influence consumer preferences—is not affected *by* market interactions, including by developing and deploying technologies.

The fact that technologies function as social structures refutes this critical assumption. Far from being structurally inconsequential, technologies participate vitally in constituting a society's basic structure. In other words, by reconfiguring our opportunities and constraints for action (including political action), influencing cultural evolution, conditioning psychological development, and in general contributing

Figure 2. *A Provisional System of Rationally-Contestable Design Criteria for Democratic Technologies and Architecture*

Toward DEMOCRATIC COMMUNITY:

A. Seek balance between communitarian/cooperative, individualized, and inter-community technologies. Avoid technologies that establish authoritarian social relationships.

Toward DEMOCRATIC WORK:

B. Seek a diverse array of flexibly-schedulable, self-actualizing technological practices. Avoid technologies that establish authoritarian social relationships.

Toward DEMOCRATIC POLITICS:

C. Insofar as technologies each tend to embody or convey symbolic meaning, avoid technologies that promote impoverished or otherwise distorted understanding.
D. Seek technologies that can help enable disadvantaged individuals and groups to participate fully in social and political life. Avoid technologies that help support illegitimately hierarchical power relations between groups, organizations, or polities.

To help secure meaningful collective self-governance:

E. Restrict the distribution of potentially adverse consequences (e.g., environmental or social harms) to within the boundaries of local political jurisdictions.
F. Seek relative local economic self-reliance.
G. Seek technologies (including an architecture of public space) compatible with egalitarian, globally-aware political decentralization and federation.

To help perpetuate democratic social structures:

H. Seek ecological sustainability.
I. Seek technologies that are relatively invulnerable to catastrophic sabotage and to the attendant risks of civil liberties abridgement.
J. Seek technologies that are compatible with: world peace, democracy abroad, and securing democratic institutions against coercion. Avoid technologies that jeopardize domestic security, provide security at the expense of democracy at home or abroad, or that contribute to international insecurity or bellicosity.
K. Seek "local" technological flexibility and "global" technological pluralism.

substantially to the texture and structure of our world, technologies do not merely—as neo-classical economists would have it—variously fail or succeed in satisfying pre-existing human wants. Rather, technologies profoundly and continuously shape the development and content of those wants, as well as the political capacity to debate and advance competing wants.[21]

In light of this insight, neo-classical economic theory and its derivative applied methodologies appear to suffer from a species of confusion and circular reasoning that, in effect, subordinates our supreme interest in evolving democratic social structures to a tail-chasing quest for efficiency in the pursuit of wants that economic institutions, interaction, and advice all help covertly to generate. In contrast, the prescriptive framework I advance takes the structural nature of technology as a starting point, and seeks to help us design and adopt technologies compatible with our being able to decide democratically what our wants social priorities should be.[22] Another way to make much the same point: where economists and other policy analysts speak of an ineluctable trade-off between democracy and efficiency, we resolve the dilemma by recognizing democracy as a pre-condition for specifying the ends with respect to which efficiency shall be judged.[23] For these reasons, I suggest that in technological decisionmaking the use of democratic design criteria should take precedence over our customary reliance on economic methodologies.[24]

But can I really be serious about privileging democracy over economics? Absolutely. For instance, one lesson of Eastern Europe's anti-Communist revolutions of 1989 is that an economy unsupervised by democracy is not only bad in itself, but also risks injustice, ecological devastation, and economic inefficiency. Or suppose that the United States' founders had decided to subject each article of the proposed Bill of Rights to a cost-benefit test? Would we be satisfied if they had rejected protection of civil liberties on the grounds that the supposed benefits could not hold a candle against the obvious economic costs?

Yet technologies function in a manner entirely comparable to a Bill of Rights. Technologies do not merely deliver sundry consumer benefits (not to mention sundry hazards and irritations); together, they constitute part of a society's core political infrastructure. Until we develop the courage to act on this insight, technological developments—whether guided by the existing market and government policies, or even by commonly espoused progressive alternatives (such as a social needs-oriented industrial policy)—will ultimately frustrate democracy, as well as the social objectives that a vigorous democracy would choose to pursue.

DEMOCRATIZING THE PROCESS OF TECHNOLOGICAL DESIGN

Having established the need to choose among alternative technologies according to their degree of substantive compatibility with democracy's structural requirements, I turn next to the question of democratic processes of technological design.

Suppose we were suddenly to achieve extensive opportunities for citizens to evaluate and choose technologies, and yet the available range of choice—that is, the variety of alternative technical means of accomplishing any given end—were quite narrow? Or suppose (as is, I suspect, generally the case today), that range included only technologies that were relatively non-democratic? In that case, the entire process of democratic participation in technological politics would remain little more than a hollow sham.

In the language of traditional electoral or legislative politics, the process of technological design (including research and development, or R&D) thus emerges as the "agenda-setting phase" in technological politics. That is, just as it is democratically essential that citizens be empowered to help shape formal legislative or electoral agendas,[25] it is likewise essential that there be extensive opportunities for citizens to participate in technological design processes. Only in this way can we be assured that when the time comes to choose among alternative technologies, the available menu of choices will afford an adequate opportunity to express our personal aspirations, view of the common good, and understanding of democracy's necessary conditions.

Broadened participation in the design process will tend to issue in a wider range of designs for several specific reasons. First, a larger number and more diverse range of participants increases the chance that someone will come up with a creative insight or breakthrough. Second, it ensures that a more diverse range of prior social needs, concerns, and experiences will be reflected in the design process. Third, it can provide enhanced opportunities for fruitful cross-fertilization of ideas from one domain of social and technological experience to another.[26]

This establishes basic and sufficient grounds for democratizing processes of technological design. But we can strengthen the case if we also take into account the resulting enhancement of opportunities for individual self-actualization and for democratic political education.

Design as Self-Actualizing Experience:
Technological design activities can potentially provide rewarding

opportunities for creativity and self-actualization.[27] And insofar as a further result of such activities is artifacts, practices, or services that go on to play a role in constituting the agenda for democratic politics of technology, then technological design can embody that special dignity which attends participation in helping to create a society's social structures.

However, experience shows that design processes can be structured so as to be either self-actualizing or else crimped and growth-impairing. In the recollection of a former machinist:

> I began . . . meeting many men whose lives revolve around the design and development of machines. I remember a dullness about them, as though the gray of the steel had entered their souls. . . . I too felt a certain lifelessness growing in me.[28]

It follows that concerted efforts are needed to design social contexts that will permit subsequent technological design efforts indeed to provide opportunities for creativity and self-expression.

Participation in Design as Political Education:
A final reason for democratizing technological design processes involves a set of accompanying political-educative benefits. First, participation in design activities can be a means of discovering the general fact, otherwise often obscured, of social contingency in technological endeavor. This is a crucial step in becoming motivated to function as a competent, critical participant in technological politics.

Second, individual involvement in design—especially within socially-open and democratically-organized settings—tends intrinsically to elicit critical reflection upon social circumstances, needs, and resources. Such reflection is vital to effective democratic politics in general, and with respect to technology in particular.

Third, from a social perspective, expanded and more diverse participation in technological design can help ensure that—via broadening of the collective perceptual field—a wider range of non-focal technological consequences will be recognized and articulated within design processes themselves, as well as in subsequent evaluation, choice and governance procedures.

Fourth, and perhaps most importantly, broadened popular involvement in design would provide citizens with democratically crucial opportunities to discover and publicize technological developments at the stage—prior to the evolution of strong institutional commitment, enduring material embodiment, uncritical accustomation, or social dependence—when there is by far the greatest potential to influence their course.[29]

Finally, each proposed or realized technological design can contribute importantly to enlarging our awareness of the specific range of functions, effects and meanings that are associated with other designs. For instance, I have found that the contrast between digital and analog wristwatches leads me to now experience the former's numeric displays as easy-to-read, but also unnecessarily precise, tending to elicit compulsiveness and a disjointed sense of time. Whereas I now consciously experience the latter—with their circular faces and cyclical displays—as connecting me, in some admittedly imperfect way, with rhythms in nature. Were citizens gradually to become aware that, whether or not it is materially realized, a publicized design alternative can thus prove educative, this would further enhance the dignity of participation in design processes. Good in itself, this would in turn also increase personal incentives to participate.

But if participatory design processes would in principle provide so many obvious democratic benefits, are citizens in practice capable of becoming competent to participate meaningfully?

Yes. Part of the reason involves recognizing that "design activity" is a complex concept, implicitly embodying a wide variety of options for different ways of participating. Moreover, the actual process of design is by no means restricted to the activity of the persons we call designers, engineers, scientists, architects, artists or craftspeople. We design when we decorate a room, create and serve a meal, or choose how to dress; workers often contribute to design in ways that go relatively unacknowledged;[30] communities engage in design when they formulate development plans or zoning laws; legislatures, courts and administrative agencies participate in design when they promulgate laws or regulations demarcating realms of permissible and impermissible technological endeavor.

For these reasons we can envision that there are many different ways in which citizens might become involved in technological design. And the range of ways in which any individual might choose, and be able effectively, to participate would increase if—as should certainly be the case—there were broader social opportunities to develop and extend one's array of design-relevant competencies. These theoretical considerations are supported empirically by case studies documenting instances of laypeople's direct participation in technological design.[31]

CONCLUSION

Building on the insight that technologies are social structures, I have argued that it is morally and democratically imperative that citizens become empowered to participate in all phases of technological politics, including evaluation, choice and governance. And I have dwelt especially on the little-heeded necessity of broadened participation in technological design, complemented by the substantive need to ensure that technologies themselves become more compatible with democracy. In the language of my introduction, the "nuts and bolts of democracy" must become more than a metaphor.

In contrast with this vision, we can interpret contemporary technological politics in terms of several interrelated components. Many of the most important decisions are actually made via a covert politics that occurs within corporate headquarters and government bureaucracies; or via the tacit politics of the economic marketplace.[32] Meanwhile, we might characterize the overt, publicly enacted politics of technology as comprising, roughly, the wrong people posing the wrong questions of the wrong technologies at the wrong time.

Think, for example, of public controversies that have engulfed the proposed fluoridation of public water supplies, siting of nuclear power plants, basing of MX missiles, office automation, or new techniques in genetic engineering.[33] Time and again we witness powerful political actors (e.g., professional politicians, government administrators, and corporate representatives), complemented by the activities of a politically unaccountable and democratically-insensitive technical elite,[34] and glorified by sensationalist, superficial, and elite-dominated press coverage,[35] ritualistically debate the economic costs and consequences, military security implications, health and safety risks, or environmental impacts of the current crop of fashionably topical technological developments. The addressed questions are not inconsequential, but this system of public discourse and decisionmaking nevertheless:

—Excludes lay citizens from anything but a trivial role;
—Often raises questions publicly only long after many of the most important decisions (e.g., over the available range of technological alternatives) have already been made elsewhere;
—Evaluates technologies almost exclusively on a case-by-case basis, thus discouraging the possibility of acting upon, or even identifying, socially-significant complementarities and synergisms among focally

unrelated technological developments;[36]
—Focuses on just a few "sexy" or "cutting edge" technologies, to the virtual exclusion of the vast majority of emerging and—even more so—existing technologies;
—Directs attention primarily toward these publicized technologies' promised material, focal purposes, at the expense of often-more-important cultural and other non-focal social consequences; and
—Leaves the question of the debated technologies' structural bearing upon the possibility of democracy as only the most important of many questions that are never asked.

So much for democratic technological procedure. Since the question of democracy is never systematically addressed, we might expect that at best contemporary technologies would prove in substantive terms to be a mixed lot: some favorable to democracy, some not, and the overall result a product of serendipity rather than autonomous human choice.

To some extent that expectation appears to be borne out. But if the design criteria listed in Figure 2 are any indication, the balance within contemporary technological orders appears actually to be skewed in decidedly anti-democratic directions.

Might that result have something to do with the erosion of participatory cultures, and of social contexts for the development and expression of citizenship? With modern tendencies toward loneliness, fragmentation of self, dissipation of purpose, narcissism, disempowerment, insecurity, and anomie?[37] With the supplanting of moral Ppand political autonomy by relative automatism?[38] With explaining the observation that in contemporary politics, "Voters elect people to be citizens. . . . But the voters are not citizens, just voters"?[39]

I suspect that the answer is, Yes. This is troubling in itself, and also implies that existing technologies will provide some hindrance against future democratization efforts—both those that are technologically-oriented, and more generally.

It will not always be easy, but we can do better. I thus uphold a vision of aspiring citizens struggling gradually to transform the technological order from an arbitrary or anti-democratic social force into a substantive constituent and expression of human freedom.

Rensselaer Polytechnic Institute

NOTES

I wish to thank Marcie Abramson, Brian Baker, Michael Black, Iain Boal, Vary Coates, David Elliott, William W. Hogan, Todd LaPorte, Laura Nader, Dick Norgaard,

Wolf Schafer, and Jeffrey Scheuer for their many helpful comments on earlier drafts of this essay. My writing was made possible, in part, through the award of a Ciriacy-Wantrup Postdoctoral Fellowship at the University of California, Berkeley.

1. See, for example, Ralph Landau and Nathan Rosenberg, eds., *The Positive Sum Strategy: Harnessing Technology for Economic Growth* (Washington, D.C.: National Academy Press, 1986).
2. See also Langdon Winner, *The Whale and the Reactor: A Search for Limits in an Age of High Technology* (Chicago: University of Chicago Press, 1986), p. 10.
3. See Richard E. Sclove, "The Nuts and Bolts of Democracy: Toward a Democratic Politics of Technological Design," in Paul T. Durbin, ed., *Critical Perspectives on Non-Academic Science and Engineering* (Bethlehem, Pa: Lehigh University Press, 1991), pp. 239-262.
4. See Susan Friend Harding, *Remaking Ibieca: Rural Life in Aragon under Franco* (Chapel Hill: University of North Carolina Press, 1984).
5. See, for example, Kenneth T. Jackson, *Crabgrass Frontier: The Suburbanization of the United States* (New York: Oxford University Press, 1985).
6. On "structure" shaping but not determining social experience, see, for example, Roberto Mangabeira Unger, *Social Theory: Its Situation and Its Task* (Cambridge: Cambridge University Press, 1987).
7. See, for example, Wiebe E. Bijker, Thomas P. Hughes, and Trevor J. Pinch, eds., *The Social Construction of Technological Systems: New Directions in the Sociology and History of Technology* (Cambridge, Mass.: MIT Press, 1987).
8. The example is from Professor Merritt Roe Smith's lectures on "Technology in America" at MIT, Fall, 1979.
9. Sherry Turkle, *The Second Self: Computers and the Human Spirit* (New York: Simon and Schuster, 1984), pp. 11-63, 324-332.
10. See A. R. Luria, *Cognitive Development: Its Cultural and Social Foundations*, ed. Michael Cole (Cambridge, Mass.: Harvard University Press, 1976).
11. See also Benjamin Barber, *Strong Democracy: Participatory Politics for a New Age* (Berkeley: University of California Press, 1984).
12. See, for example, K. Guild Nichols, *Technology on Trial: Public Participation in Decision-Making Related to Science and Technology* (Paris: Organisation for Economic Cooperation and Development, 1979).
13. I explain and justify these conditions in Richard E. Sclove, *Technology and Freedom: Toward a Democratic Politics of Technology, Architecture, and Design* (forthcoming), chapters 3 and 4. Suppose one is sympathetic to the basic notion of an egalitarian, participatory democracy, but dissatisfied with my particular characterization of democracy's necessary conditions? That would suggest accepting my general theory of democracy and technology, while perhaps wishing to seek an alternative set of design criteria for democratic technologies—that is, a set compatible with a different specification of democratic institutions.
14. Quoted in Gail Bass, *et al.*, *The Appeal of Nuclear Crimes to the Spectrum of Potential Adversaries* (R-2803-SL) (Santa Monica: RAND Corp., Feb. 1982), p. 19.
15. Compare Russell W. Ayres, "Policing Plutonium: The Civil Liberties Fallout," *Harvard Civil Rights-Civil Liberties Law Review* 10 (Spring 1975): 369-443.
16. See, for example, Amory B. Lovins and L. Hunter Lovins, *Brittle Power: Energy Strategy for National Security* (Andover, Mass.: Brick House, 1982).
17. See, for example, Dolores Hayden, "Capitalism, Socialism and the Built Environment," in Steve Rosskam Shalom, ed., *Socialist Visions* (Boston: South End Press, 1983), pp. 59-81.
18. Dolores Hayden, *Redesigning the American Dream: The Future of Housing, Work, and Family Life* (New York: Norton, 1984), p. 189.

19. Robert Schrank, *Ten Thousand Working Days* (Cambridge, Mass.: MIT Press, 1978), p. 226.
20. For the derivation of these criteria, see Sclove, *Technology and Freedom* (note 13, above), Part II.
21. This is a point overlooked in K. S. Shrader-Frechette's attempt to critically defend cost-risk-benefit analysis in *Science Policy, Ethics, and Economic Methodology: Some Problems of Technology Assessment and Environmental-Impact Analysis* (Dordrecht: Reidel, 1985). On the other hand, by according priority to the development of democratic social structures, the alternative conceptual framework that I propose appears to satisfy Shrader-Frechette's suggestion in chapter 8 that evaluative systems be "lexicographically ordered." See also note 23, below.
22. I elaborate these points in Sclove, *Technology and Freedom*, chapter 19.
23. In many cases democratization actually enhances economic efficiency, as indicated by such measures as labor productivity. See, for example, Samuel Bowles, David M. Gordon, and Thomas E. Weisskopf, *Beyond the Wasteland: A Democratic Alternative to Economic Decline* (Garden City, N.Y.: Anchor/Doubleday, 1983). Nevertheless, a stronger moral point is that our highest-order interest in democracy must never be sacrificed on the altar of empirical contingency. See John Rawls, *A Theory of Justice* (Cambridge, Mass.: Harvard University Press, 1971), on the lexical priority of equal liberty over his "difference principle." Thus, even when structural democratization might prove economically inefficient, we should proceed with it—subject to the qualification that appears in the following note.
24. The one circumstance in which we ought to treat economic considerations on a par with democracy is when economic circumstances themselves develop structural implications for the possibility of democracy. The obvious case is an economy functioning so clumsily that it cannot meet citizens' needs for health and survival. Biological survival, it goes without saying, is a necessary condition of democracy. In such cases, indeed economic considerations should be taken into account on a high priority basis. But only temporarily—that is, only as long as it takes to mitigate the justifying circumstances; and still without presumptively preempting other considerations that bear on democracy's possibility.
25. See, for example, Roger W. Cobb and Charles D. Elder, *Participation in American Politics: The Dynamics of Agenda-Building* (Baltimore: Johns Hopkins University Press, 1972).
26. Several historical examples of such cross-fertilization are documented in Dolores Hayden, *Seven American Utopias: The Architecture of Communitarian Socialism, 1790-1975* (Cambridge, Mass.: MIT Press, 1976), pp. 197-198, 355.
27. On the democratically essential role of self-actualization, see Figures 1 and 2, above. For examples of design as a self-actualizing activity, see Sclove, *Technology and Freedom*, chapter 20.
28. Schrank, *Ten Thousand Working Days*, p. 135.
29. On the loss of flexibility as a technology's development or deployment continues, see David Collingridge, *The Social Control of Technology* (London: Frances Pinter, 1980).
30. See Tony Manwaring and Stephen Wood, "The Ghost in the Machine: Tacit Skills in the Labor Process," *Socialist Review* 14:2 (March-April 1984): 57-83 and 94.
31. See Sclove, "The Nuts and Bolts of Democracy: Toward a Democratic Politics of Technological Design" (note 3, above).
32. See, for example, Bijker, Hughes, and Pinch, *Social Construction of Technological Systems*.
33. See, for example, Dorothy Nelkin, ed., *Controversy: Politics of Technical Decisions*, 2d ed. (Beverly Hills: Sage, 1984).
34. See, for example, Richard E. Sclove, "Energy Policy and Democratic Theory,"

in Dorothy S. Zinberg, ed., *Uncertain Power: The Struggle for a National Energy Policy* (New York: Pergamon, 1983), pp. 41-48.

35. Dorothy Nelkin, *Science, Technology and the Press* (New York: Freeman, 1987).

36. For historical examples of such synergisms, see, for example, Michel Foucault, *Power/Knowledge: Selected Interviews and Other Writings, 1972-77*, ed. Colin Gordon (New York: Pantheon, 1980); and Albert Borgmann, *Technology and the Character of Contemporary Life* (Chicago: University of Chicago Press, 1984).

37. For documentation and discussion of these ills, see, for example, Borgmann, *Technology and Contemporary Life*, and Robert N. Bellah, et al., *Habits of the Heart: Individualism and Commitment in American Life* (Berkeley: University of California Press, 1985).

38. See, for example, Robert Kegan, *The Evolving Self: Problem and Process in Human Development* (Cambridge, Mass.: Harvard University Press, 1982), pp. 207-215, 243-249; and Jacob Needleman, *Consciousness and Tradition* (New York: Crossroad, 1982), pp. 72-87.

39. Karl Hess, "The Politics of Place," *Co-Evolution Quarterly*, no. 30 (Summer 1981), p. 6.

PART III

HISTORICAL AND CULTURAL REFLECTIONS

STANLEY R. CARPENTER

INSTRUMENTALISTS AND EXPRESSIVISTS: AMBIGUOUS LINKS BETWEEN TECHNOLOGY AND DEMOCRACY

Reflecting on his visit to America, Alexis de Tocqueville worried that the democratic taste for freedom, which he found here, would prove to be weaker than the democratic drive toward equality.[1] To a degree he could not have imagined, that democratic drive has been progressively embodied in technology. It thus appears that de Tocqueville's predictions concerning equality have been realized, in that, to a significant degree, ours is a democracy of common possessions, egalitarianism expressed technologically. But it is also possible to make the case that the drive toward unlimited production and indefinitely expanding modes of consumption that have characterized this technological action, have resulted in a progressive loss of freedom to adapt to an increasingly overstressed biosphere. I argue here that an inability to design more appropriate technologies is reflective of an ambiguity in understanding technology itself. Two strands of moral discourse, that lead to conflicting images of technology are identified here. One, traceable to Enlightenment themes, embodies political atomism and leads to an instrumental conception of technology. The second, reflective of subsequent Romantic ideas, gives credence to the claim that technology is a mode of expression, a self-defining way of being in the world. Our understanding of technology should move beyond this ambiguity. By identifying some of its roots, I hope to begin to develop a new synthesis of understanding about technology, one displaying a clearer appreciation of the biospheric limits that constrain technological action.

INTRODUCTION

A community acquires self-understanding by means of its characteristic conversations. Recurring, often disparate, themes amount to constitutive moral languages which reflexively define the group, locating it within a web of traditions, and keeping alive strains of dissonance. In the best of cases, such conversations comprise sustained and constructive arguments in which the dissonances contribute to a healthy societal pluralism, a true *communitas communitatum*. At the other extreme, when the arguments degenerate, when differences become mutually exclusive, when there ceases to be consensus on the nature of the good life or what it means to be fully human, we may begin to speak of a "crisis of legitimation."[2]

In the face of such crises, our behavior tends to be self-destructive. We act so as to weaken and eventually destroy the conditions which make our actions possible. Thomas Merton, for example, notes the irony of the American pioneer who was both defined by the frontier and relentless in pursuit of its elimination.[3] Garrett Hardin identifies as the epitome of tragedy the combination of free-market forces and common resources. Left alone, the process leads inexorably to the eventual destruction of the common resource itself.[4]

Loss of legitimacy can occur, as Charles Taylor shows,[5] when a practice becomes such an entrenched tradition that it is taken for granted. In this case alternative formulations seem inconceivable. As a practice sinks to the level of common sense, the controversial claims, qualifications, unresolved problems, and explicit exceptions that were clearly evident at its inception, seem increasingly beyond reach. When times change, the practice cannot. Ideology becomes little more than encrusted dogma.

In addition, the identities of group, community, and nation can lose their center if the moral languages that contribute to that identity lose vitality or become overpowering and intolerant of competing ideas. Thus, de Tocqueville claimed that the destructive forces of radical individualism, which had been so much in evidence in France, had up to the time of his visit been counterbalanced in the United States by biblical communitarianism and the strong national habit of forming voluntary associations.[6] As the nineteenth century passed the quarter mark, however, he worried that the forces of localism and community were weakening in the face of inevitable, encroaching mass democracy.

What concerned de Tocqueville was the drift of democracy. He worried that the dynamic energies released in democratic action would continue to push toward a collective uniformity; equality would be won at the price of freedom. He wrote: "I think that democratic communities have a natural taste for freedom: left to themselves, they will seek it, cherish it and view any privation of it with regret. But for equality, their passion is ardent, incessant, invincible: they call for equality in freedom; and if they cannot obtain that, they will call for equality in slavery."[7]

Fifty years later, Lord Bryce would observe that de Tocqueville's dire predictions had not materialized. "Faint are the traces which remain of that intolerance of heterodoxy in politics or social views whereon he dilates."[8] Yet, we continue to gaze with fascination at the picture of ourselves painted by de Tocqueville. One senses the impressive degree to which he got us Americans right. We are even ready to grant him a certain prescience concerning the eventual convergence of social forms into monolithic mass society, the realization, that is, of his fears about the end of freedom. We *are* a mass democracy, a democracy of common possessions, a democracy made possible by the broadest possible distribution of consumer items and services of every description. This democracy, to an unprecedented degree, is embodied in technology. Our political impulses are *expressed* by means of technological acts and mass produced artifacts.[9] Yet these same egalitarian impulses, expressed in consumptive lifestyles, can render us unfree and politically impotent if we are unable to abandon them despite clear evidence that they are no longer appropriate.

To link technology thus directly with politics is to follow one strand of moral discourse that has helped to define us. In this light, technology can profitably be understood as an expressive language of social action.[10] It is more common, however, to discuss technology in purely instrumental terms: as a means to separately generated ends. This approach reflects an alternative strand of moral discourse, one that deserves our attention because it fosters the commonly held assumption that technology, being merely instrumental, is also morally neutral.

This, then, is one difficulty with attempting to link technology and democracy. On the one hand, the relationship seems constitutive. On the other, technology and democracy have weak, idiosyncratic ties. There is plenty at stake here. Technology is itself changing in such ominous ways that the old terms of past conversations, and old assumptions cited in their support, need reexamination. Novel technologies which tamper with the genetic code of life have qualitatively changed technological action and introduced new moral imperatives.[11] The global scale of modern technologies has transformed international politics into global politics and has generated an entire class of supranational environmental issues.[12] In view of such developments, the prevailing understandings and assessments of technology need to be revised. What I offer here is an interpretation of two strands of thought that have characterized images of technology and politics in modern thought. My discussion aims to identify and historically situate problems of ambivalence and even incommensurability that plague present-day assessments and policy recommendations about the direction of technological change.

INSTRUMENTALISTS AND EXPRESSIVISTS

Taylor argues that modernity is dichotomous. Two complex patterns of beliefs and values, two cultural strands, two ontologies, coexist in dialectical tension. On the one hand, we are children of the Enlightenment; on the other, heirs of the Romantics.[13] One line of thought traces its heritage to Descartes, Bacon, Hobbes, and Locke, while the other bears the stamp of Rousseau, Herder, Humboldt, Marx, and Bradley, among others. Taylor surrounds core identifications of Enlightenment and Romantic themes with a rich variety of disparate themes deftly woven into two identifiable tapestries.[14] We are the heirs of both traditions.

Western philosophy in the twentieth century has had much to say about intentionality. Frege, Russell, the Wittgenstein of the *Tractatus*, Carnap, for example, hold that intentionality of thought is logically prior to that of speech. Furthermore, they defend atomistic semantics; there are independent atoms of meaning. Dewey, Heidegger, and the Wittgenstein of the *Investigations* regard intentionality of speech as logically prior to that of thought. What is more, in opposition to foundationalists, they hold that meanings are always culturally embedded, both diachronically and synchronically.[15]

As he explores the first of these positions, Taylor notes: "On this view, language is a set of instruments which lie, as it were, transparently to hand and which can be used to marshall ideas, this use being something we can fully control and oversee."[16] For my purposes here I will call this view the "instrumentalist" position. A strongly contrasting position can be found in the nineteenth-century writings about language of Herder and Humboldt, for example. It conceives of language not as a set of separate instruments, but "in the nature of a web.... Because the words we use now only have sense through their place in the whole web, we can never in principle have a clear oversight.... Our language is always more than we can encompass; it is in a sense inexhaustible."[17] This orientation to language Taylor calls "expressive," the view that, "We express/realize a certain way of being in the world, that of reflective awareness."[18] In the following discussion I shall refer to this as the "expressivist" position.

When applied to the task of describing technology, Enlightenment instrumentalism produces a strikingly different picture. In the first instance, both language and technology function instrumentally: in the case of language, to mirror epistemologically prior ideas and meanings; and for technology, to provide an ensemble of instruments for the satisfaction of independently generated individual preferences. Regarded expressively, however, language and technology are not sharply distinguishable. Technology as social *praxis* expresses a way of being in the world; but so do other symbolic achievements such as prose or poetic instances of natural languages and artistic creations such as music, dance, and painting.

To help understand the relationship between technology and politics, we need to explore how instrumentalist and expressivist views deal with the nature of the individual, the individual's relation to the political order, and the role of technology within that order. According to the instrumentalist position, individuality must be understood as ontologically prior to community. In the tradition of Hobbes and Locke, one starts with an abstract, punctual self, defined as a locus of desires, needs, purposes, fears. This self is a disengaged atom, a bearer of preferences and of rights. Hence the recognized task of the theories of social contract of classical liberalism is to devise social arrangements that address needs and preferences of atomistic selves.[19] Such arrangements are extrinsic, experimental, and thus provisional. Their purposes are utilitarian. Whereas societies manifest convergence of purpose toward political, economic, scientific, technological, even "spiritual" goals, one must never forget that such societies are always collections of "independent and rational beings, who are the sole generators of their own wants and preferences, and the best judges of their own interests."[20]

Picturing the self as atomistic, disengaged, and ontologically fundamental, with society as an accidental collocation, was a distinctly modern idea at odds with traditional understandings of humans as *homo politicus*.[21] The nineteenth century, however, brought strong challenges to this atomistic conception of modernity. Stephen Lukes identifies several sources: Romantic Conservatism in France, England, Germany; Hegel and Marx and their followers; positivists such as Comte; sociologists such as Durkheim; English Idealists, such as T. H. Green, and American Pragmatists, especially Mead.[22] Marx condenses this viewpoint in his dictum, "It is not the consciousness of men that determines their being, but, on the contrary, their social being that determines their consciousness."[23]

Fundamental to expressivist theories is the claim that society defines the individual. F. H. Bradley, for example, makes clear his opposition to atomistic theories of individualism in outspoken terms: "Man is a social being; he is real only because he is social, and can realize himself only because it is as social that he realizes himself. The mere individual is a delusion of theory; and the attempt to realize it in practice is the starvation and mutilation of human nature, with total sterility or the production of monstrosities."[24]

Taylor mounts a comparable challenge to what he calls the "thin conception of the self" in the following terms:

> The basic error of atomism in all its forms is that it fails to take account of the degree to which the free individual with his own goals and aspirations, whose just rewards it is trying to protect, is himself only possible within a certain kind of civilization; that it took a long development of certain institutions and practices, of the rule of law, of rules of equal respect, of habits of common deliberation, of common association, of cultural self-development, and so on to produce the modern individualist; and that without these the very sense of oneself as an individual in the modern meaning of the term would atrophy.[25]

Just as expressivist theories of language locate intentionality in the self-expression of communities, so the counterpart expressivist theory of self regards human individuality as emerging from a social background of mutual understandings, evaluations, and proscriptions. The relation of the individual to this background is complex. Rousseau, for instance, holds that an authentic relationship between the individual and the state, as expressed by the General Will, is capable of morally transforming and redefining the individual. A new person is created when the corrupting influences of intermediate allegiances are removed. Marx, by contrast, holds that every social order that retains the repressive mechanism of the state is necessarily corrupt.

In the instrumentalist view the atomistic self is a locus of individualistic desires, preferences, fears, and rights. In contrast, the expressivist position regards the self as an intersection of social relationships. For instrumentalists the relationship between the individual and the society is wholly extrinsic. Dimensions of social life are constructed, as it were, out of externally related utilitarian categories: the political, the economic, the scientific, the technological. Expressivists find it much harder to make these distinctions. They regard life as all of a piece, something better pictured as a scenario rather than separable into sociological categories.

It is not surprising that such different conceptions of the self and society lead to incompatible models of technology. Because we are heirs to both Enlightenment and Romantic traditions, we are unable to make a clean break with either of them. For that reason, discussions about the meaning of technology in our lives today are unavoidably ambiguous.

An instrumentalist model of human nature is stark in its ontological isolation of the individual. The punctual self is a disengaged bearer of preferences. The physical and social world are built up of independent components. One engages the world in politics, economics, science, technology, religion, philosophy. Each constitutes a distinct social invention for focusing the interests of the individual. There is no particular thematic unity to these institutions. They exist because they meet a particular need in a given historical era. The utility of medieval religion vanishes, for example, in the modern age when naturalism strips enchantment from everyday life.[26] Conversely, economics emerges as a separate social institution, as Karl Polanyi sees it, when the medieval era is replaced by one of commerce and industry.[27] One may speak of the political dimensions of technology or the technological aspects of politics, or again of the economic dimensions of either. But one runs the risk of widespread misunderstanding if one argues that there are cases in which drawing a firm line between politics and technology makes no sense. Winner's analyses of inherently political technologies seem downright perplexing according to an instrumentalist approach to social analysis.[28] For, surely, political and technological affairs exist in separate realms.

The contemporary method of public policy formulation known as Technology Assessment (TA)[29] adopts the instrumentalist approach. To begin to deal with the future impacts of technological change, the problem is broken down into separate categories: political issues are isolated from economic ones; scientific matters are differentiated from technological concerns; environmental and legal factors are addressed separately. Because TAs are almost always performed by interdisciplinary teams, with a representative of each of several disciplines responsible for one section of the final report, the piecemeal quality of a TA is fully evident.

Expressivist theories of individuality consider abstract individualism to be an incoherent idea. "It is impossible," observes Marxist scholar Leszek Kolakowski, "for man to conceive of himself in his independence from nature, or of nature in its independence from his practical contact with it."[30] Rather than describing such social organizations as community, guild, market, as provisional and instrumental collections of social atoms, Kolakowski observes, "The categories into which this world has been divided are not the result of convention or a conscious social agreement; instead they are created by a spontaneous endeavor to conquer the opposition of things. It is this effort to subdue the chaos of reality that defines not only the history of mankind, but also the history of nature as an object of human needs—and we are capable of comprehending it only in this form."[31] This spontaneous endeavor, which Marx called "sensuous human activity," is nothing other than consciousness itself. Consciousness is thus not a metaphysically prior atom of individuality in the Cartesian sense, but rather an expression of historically determined social relationships. Technological artifacts, are, on this account, congealed forms of these relationships.

Marx emphasized the historically disparate forms which "sensuous human activity" or *praxis* could assume. Throughout the tortuous history of *praxis*, technology was often expressive of hostile political arrangements and alienating and self-canceling human efforts. In some readings, Marx seems to argue that *praxis* can take the form of free labor heroically struggling against existing power relationships.[32]

The recent work of Joseph Margolis draws upon both Marx and Heidegger to reformulate an expressivist approach to technology. He writes,

> The technological . . . performs a double role. On the one hand, in accord with Heidegger's and Marx's view, it signifies how reality is 'disclosed' to humans—primarily because it is through social production and attention to the conditions of survival (both precognitively and through explicit inquiry) that our sense of being in touch with reality is vindicated at all; but contrary to the thrust of Heidegger's later qualification, the correction of all theories of cognition and reality thus informed is itself inevitably historicized and subject to the ideological limits of any successor stage of *praxis*.[33]

As I read him, Margolis invests technology with the power of disclosure of the world itself. But he claims this disclosure must always contain historically contingent, ideological elements. Artifacts disclose the world to us, but the disclosure is never free from ideological bias. Subjective and objective elements are inextricably linked. Artifacts thus speak in a language that is temporally contingent and developmental. As David Dickson observes:

> Technology can be considered as a "language" of social action. From this language we select those individual elements—whether tools or machines—which we then apply to carry out a particular task. The instrumental (means to an end) activity of the individual, carried out through the successive applications of a variety of machines, might be considered as a personalized form of "speech." . . . And in the way that language is experienced through the spoken or written word, which therefore become the "objectifications of social ideas and concepts," so, as we have seen above, technology represents in the individual machines of which it is composed the "objectifications" of social action.[34]

So characterized, there is a cogency to claims that technology performs a rhetorical function, as when we express ourselves by means of disposably constituted technological environments while simultaneously bemoaning the impermanence and ever quickening pace of modern life. When, on the other hand, technology is taken to be an accidental collection of useful instruments, available for carrying out independently generated life plans, its sheer instrumentality renders it innocuous and evaluatively neutral.

An ambiguity in our understanding of technology, the gap between instrumentalist and expressivist understandings, afflicts many contemporary efforts to evaluate the links between technology and the political order. Instrumentally, technology simply reflects human intentions; expressively, it constitutes them. Instrumentally, technology enables us to add new bits of furniture to our world; expressively, technology designs that world and gives form to our most basic priorities and necessities. Instrumentally, technology serves to meet independently generated preference and need schedules. Expressively, technology actually defines those needs, providing a basis for our self-understanding.

From the standpoint I have sketched, we can reinterpret de Tocqueville's qualms about the relationship between equality and freedom in modern democracy. Today's society may be described as a democracy of common possessions, a technology-constituted democracy in which the drive toward unlimited production and infinitely expanding consumption produces the outcome de Tocqueville feared—equality affirmed at the expense of human freedom. How does this phenomenon appear from instrumentalist and expressivist perspectives?

Understood instrumentally, growth-oriented production processes can be shown to be objectively efficient. Design norms promoting technical efficiency are morally neutral, independent of and clearly separable from political questions about freedom or equality.[35] The instrumentalist view of consumption is also morally neutral. Technology is truly protean, servant, not master, awaiting our beck and call with no hint of approbation or criticism.

From an expressivist viewpoint, growth-oriented production technologies communicate the message that this world is an unlimited source of raw materials and disposal sites for worn-out artifacts and spent materials. As consumer items, technologies serve as fetishes of power and status, letting us "do our thing," be ourselves. The technologies constitute us, making us go fast, be virile, or turn us into "couch potatoes" or "hedonistic globules of desire."[36]

De Tocqueville claimed that democracies tend to favor equality over freedom. By our choice of a technologically constituted democracy—a democracy of common possessions—realized against background assumptions and economic models of unlimited production and infinitely expanding modes of consumption, de Tocqueville's concerns over the diminution in freedom have been realized. This technological path is a dead end, but we appear unable to abandon it.

What, we may ask, of de Tocqueville's assertion that democracy ends up causing one to forget other generations, past and future? Here too he was correct. Technologically expressed democracies do this. Alvin Toffler has coined the phrase "future shock" to explain the forward momentum of technological societies and the psychological inability of their citizens to adjust to the pace of change they produce.[37] For us moderns the term "new" is an encomium. Even in the midst of our bewilderment and shock at the pace of change, we proceed to obliterate outmoded artifacts and practices to make room for "revolutionary" changes.[38] At my highly computerized campus library, for example, books are endangered artifacts.

Because social *praxis* is expressive of growth-based modes of production, it is as negligent of future generations as it is forgetful of a technically obsolete past. Neglect of the long-term future is justified by the discount rate of money projected forward as little as one generation. Depletion of scarce resources is countered by an unshakable faith that technological innovation will produce substitute materials or processes in a timely manner.

What are the prospects for a synthesis of instrumentalist and expressivist understandings of technology? Some proposals suggest a thoroughgoing change of mind. We need, it is argued, a new way of thinking, a way characterized by a new naturalism, a new holism, and a new immanentism.[39] The new naturalism embodies a semi-deterministic monism. New holism places an emphasis on systemic models of Nature and discounts the explanatory efficacy of reductive materialism. The new immanentism seeks sources of value within Nature and not "outside" it.

Others believe that it is possible for us to extricate ourselves sufficiently from our technologies to refashion them in ways more compatible with biospheric limits. Could we have a "tech-fix" transition toward a new world order? As I have discussed elsewhere,[40] the 1970s saw numerous attempts to construct "alternative" or "appropriate" technologies that were regionally sensitive, and ecologically gentle, technologies that fostered a maximum of democracy and sustainability. While these approaches were often technologically clever, they were psychologically and politically naive. An ignorance of the history of technology, accompanied, in some cases, by an anti-intellectualist bias, led proponents of alternative technologies to underestimate or misconceive the complex process of producing technological change. For the reasons I have identified, addressing this ignorance will not be straightforward or simple.

CONCLUSION

Today's attempts to develop a prescription for a democratic, sustainable technology are hampered by a lack of clarity in ideas about the nature of technology itself. An unresolved tension between Enlightenment instrumentalism and Romantic expressivism is lodged "in our bones." Thus, even well-meaning efforts to correct past technological abuses are threatened by confusions in our self-understanding. My diagnosis has pointed to some of the roots of this predicament. It is only by continued conversation about the nature of technology and its constitutive role in the good life that glimmerings of a new synthesis between instrumental and expressive modes of understanding may begin to emerge.[41]

Georgia Institute of Technology

NOTES

1. Alexis de Tocqueville, *Democracy in America*, ed. R. D. Heffner, 2 vols. (New York: Vintage, 1954).
2. As is well known, Habermas uses this phrase to identify contradictions in the modern advanced capitalist societies; see his *Legitimation Crisis* (Boston: Beacon Press, 1975). I take it herein in a more general sense as defined by Charles Taylor to mean, "Societies destroy themselves when they violate the conditions of legitimacy which they themselves tend to posit and inculcate"; see Charles Taylor, in "Legitimation Crisis," in *Philosophy and the Human Sciences: Philosophical Papers* (Cambridge: Cambridge University Press, 1985), vol. 2, p. 248.
3. Thomas Merton, "The Wild Places," *The Center Magazine* (July 1968), pp. 40-44.
4. Garrett Hardin, "The Tragedy of the Commons," *Science* 162 (13 December 1968): 1243-1248.
5. Charles Taylor, "Philosophy and Its History," in R. Rorty, J. Schneewind, and Q. Skinner, eds., *Philosophy in History* (Cambridge: Cambridge University Press, 1984), pp. 17-30; see discussion of this historical approach in Stanley R. Carpenter, "What Technologies Transfer: The Contingent Nature of Cultural Responses," in J. Pitt and E. Byrne, eds., *Technological Transformation: Contextual and Conceptual Implications* (Dordrecht: Kluwer, 1989), pp. 163-177.
6. The weakening of these counterbalancing forces, and a ruthless realization of the ontological isolationism of abstract individualism, is the concern of Robert N. Bellah, et al., *Habits of the Heart: Individualism and Commitment in American Life* (Berkeley: University of California Press, 1985).
7. De Tocqueville, *Democracy in America*, p. 192.
8. Cited by Richard D. Heffner, in editor's introduction to *Democracy in Action*, p. 21.
9. This is a central theme of both of Winner's books, *The Whale and the Reactor* (Chicago: University of Chicago Press, 1986), esp. pp. 3-18; and his earlier *Autonomous Technology: Tecnics-Out-of-Control as a Theme in Political Thought* (Cambridge, Mass.: MIT Press, 1977). See also Borgmann's defense of the thesis that the freedom that classical liberalism claims to leave open is in fact closed by the technological configurations of modern industrial societies, in Albert Borgmann, *Technology and the Character of Contemporary Life* (Chicago: University of Chicago Press, 1984).
10. See Carpenter, "What Technologies Transfer," section entitled "Technology as Language."
11. One of the first to emphasize this new ethical imperative is Hans Jonas, *Imperatives of Responsibility: In Search of a New Ethics for the Technological Age*

(Chicago: University of Chicago Press, 1984). This is a translation by Jonas himself, in collaboration with David Herr, of *Das Prinzip Verantwortung: Versuch einer Ethik für die technologische Zivilisation*, and *Macht oder Ohnmacht der Subjektivität? Das Leib-Seele-Problem im Vorfeld des Prinzips Verantwortung* (Frankfurt: Insel Verlag, 1979).

12. In a famous speech of 2 November 1987, Soviet General Secretary Mikhail S. Gorbachev noted that the modern world is "mutually connected, mutually dependent, and forms a definite, integral whole. This is brought about by the internalization of world economic links, the all-embracing nature of the scientific-technical revolution, the fundamentally new role of the media and communications, the condition of the planet's resources, the overall ecological danger, and the glaring social problems of the developing world which affect everyone. But mainly it is brought about by the rise of the problem of mankind's survival, for the appearance and the threat of the use of nuclear weapons has placed a question mark over (mankind's) very existence."

13. See esp. Charles Taylor, *Human Agency and Language: Philosophical Papers I* (Cambridge: Cambridge University Press, 1985), pp. 15-44, and *Philosophy and the Human Sciences: Philosophical Papers II* (Cambridge: Cambridge University Press, 1985), pp. 187-337.

14. In addition to the terms "enlightenment" and "romantic," Taylor also adopts the terms "designative" and "expressive" to refer to his core themes. See vol. I, pp. 217-219, 246 and 247. My use herein of "instrumentalist" and "expressivist" is closest to this latter stipulation.

15. A summary of the dichotomous positions regarding "intentionality" in twentieth-century Western philosophy is provided in *APA Newsletter on the Teaching of Philosophy* 2:3 (1981): 2.

16. Taylor, *Philosophical Papers I*, p. 217.

17. *Ibid.*, p. 231.

18. *Ibid.*, p. 232.

19. For concise discussions of abstract individualism, as well as political, economic, religious, ethical, epistemological, and methodological individualism, see Stephen Lukes, *Individualism* (Oxford: Blackwell, 1973), esp. pp. 73-79.

20. *Ibid.*, p. 79.

21. See discussions of this point by C. B. Macpherson, in *The Political Theory of Possessive Individualism* (Oxford: Clarendon Press, 1964), introduction; also Victor Ferkiss, *The Future of Technological Civilization* (New York: Braziller, 1974), p. 23.

22. Lukes, *Individualism*, p. 78.

23. Karl Marx, *Living Thoughts of Karl Marx* (New York: Longmans, Green, 1939).

24. F. H. Bradley, "My Station and its Duties," in his *Ethical Studies*, 2d ed. (Oxford: Oxford University Press, 1927), p. 174.

25. Taylor, *Philosophical Papers II*, p. 309.

26. This is of course how Max Weber characterized a mechanistic reorientation to the world exemplified by the Puritans. The term for disenchantment is *Entzauberung*. See an argument for a reversal of this stance in David Ray Griffin, "Modern Science and the Disenchantment of the World," in D. Griffin, ed., *The Reenchantment of Science* (Albany: State University of New York Press, 1988), pp. 2-21.

27. Karl Polanyi, *The Great Transformation* (New York: Rinehart, 1944), p. 71.

28. See Winner, "Do Artifacts Have Politics?" in *The Whale and the Reactor*, pp. 29-39.

29. A. Porter, F. Rossini, S. Carpenter, T. Roper, *A Guidebook for Technology Assessment and Impact Analysis* (New York: North Holland, 1980). See also Stanley R. Carpenter, "Technoaxiology: Appropriate Norms for Technology Assessment," in P. Durbin and F. Rapp, eds., *Philosophy and Technology* (Dordrecht: Reidel, 1983), pp. 115-136.

30. Leszek Kolakowski, "Karl Marx and the Classical Definition of Truth," in *Marxism and Beyond: On Historical Understanding and Individual Responsibility*, trans. Jane Z. Peel (London: Pall Mall Press, 1968), pp.64 and 65.

31. *Ibid.*, p. 66.

32. Alvin Gouldner says, "Marx has two tacitly different conceptions of praxis . . .: Praxis$_1$ is the unreflective labor on which capitalism rests, the wage labor imposed by necessity which operates within its confining property institutions and its stunting divisions of labor. While this labor inflicts an alienation upon workers, it also constitutes the foundation of that society, reproducing the very limits crippling workers. Here workers are constrained to contribute to the very system that alienates them. This conception of praxis is congenial to Scientific Marxism. In the second, more heroic concept of practice—Praxis$_2$, more congenial to Critical Marxism—emphasis is on a practice that is more freely chosen, more especially on political struggle. If Praxis$_1$ is the constrained labor that reproduces the status quo, Praxis$_2$ is the free labor contributing toward emancipation from it. In undertaking the first form of labor or practice, persons submit to necessity; in the second, however, they undertake a deliberate and Promethean struggle against it." See his *The Two Marxisms* (New York: Oxford University Press, 1980), pp. 33-34; cited by Joseph Margolis in "Three Conceptions of Technology: Satanic, Titanic, Human," in P. Durbin, ed., *Research in Philosophy and Technology*, vol. 7 (Greenwich, Conn.: JAI Press, 1984), p. 147.

33. Joseph Margolis, "Pragmatism, Transcendental Arguments, and the Technological," in Durbin and Rapp, *Philosophy and Technology*, pp. 305-306.

34. David Dickson, *The Politics of Alternative Technology* (New York: Universe Books, 1974), pp. 176-177.

35. See discussion of the concept of "efficiency" in Stanley R. Carpenter, "Alternative Technology and the Norm of Efficiency," in P. Durbin, ed., *Research in Philosophy and Technology*, vol. 6 (Greenwich, Conn.: JAI Press, 1983), pp. 65-76.

36. Thorstein Veblen, "Why is Economics not an Evolutionary Science?" in *The Place of Science in Modern Civilization and Other Essays* (New York: Huebsch, 1919), p. 73.

37. Alvin Toffler, *Future Shock* (New York: Bantam, 1970).

38. See Winner's engaging discussion of the word magic surrounding the application of the term "revolutionary" to both trivial and major instances of technological change in "Mythinformation," in *Whale and Reactor*, pp. 98-117.

39. Victor Ferkiss, *Technological Man: The Myth and the Reality* (New York: Mentor, 1969), pp. 202-223; and *The Future of Technological Civilization* (New York: Braziller, 1974), esp. pp. 213-293. The early chapters of the latter book are a sustained attack on the political philosophy of classical liberalism. See the comparable approach by Macpherson in *The Political Theory of Possessive Individualism*.

40. Stanley R. Carpenter, "A Conversation concerning Technology: The Appropriate Technology Movement," in P. Durbin, ed., *Technology and Contemporary Life* (Dordrecht: Reidel, 1978), pp. 87-106.

41. An earlier draft of this paper was presented at the fifth biennial international conference of the Society for Philosophy and Technology, July 1, 1989, at the University of Bordeaux, Bordeaux, France. The author wishes to thank Langdon Winner for many helpful suggestions in the preparation of this revised draft.

VITALY GOROKHOV

POLITICS, PROGRESS, AND ENGINEERING: TECHNICAL PROFESSIONALS IN RUSSIA

In Russia nowadays we are increasingly concerned to examine our history and the recent past. We hope to restore truths that have been forgotten as well as ones intentionally concealed. Many of us understand that the moral improvement of our society cannot be achieved without open, continuing reappraisal of links between the present and what came before. Our history offers patterns that are worth reviving. It is essential to understand why these patterns disappeared and to ponder the lessons that might yet be drawn from them. This can help us overcome the negative influences that still exist in our society today.

The history I shall outline is that of Russian engineering. The first professional associations of engineers appeared during the nineteenth century in a number of industrial countries, especially England. The first of these were training clubs set up as professional communities of engineers in different fields, for instance, the Institute of Civil Engineers and the Society of Ship-Building Engineers in England. These were followed by nationwide trade unions embracing engineers in a wide variety of fields. Among the first of this kind was the Union of German Engineers established in 1856. The books and magazines published by the union and the research it carried out were of great interest to Russian engineers during this formative period.

The Russian Technical Society was founded in 1866; the title Imperial was added to its name in 1874. According to its charter, the RTS was set up to promote technology and technological industries in Russia through:

1) readings, meetings, and public lectures on technical subjects;

2) disseminating theoretical and practical information with the help of periodicals and other publications;

3) expanding technical education;

4) suggestions for tackling technological problems facing the national industries, with bonuses and medals awarded for the best ways to resolve them;

5) exhibitions of manufactured goods and technical products;

6) studies of factory materials, products, and methods and techniques of work most widespread in Russia, both on the Society's choice and on requests from other societies and private persons;

7) founding a technical library, chemical laboratory, and technical museum;

8) mediation between technicians and persons in need of their services;

9) aiding the sales of products by indigenous craft industries;

10) appealing to the government for measures that could serve to advance the technical industries.

In 1877, a Polytechnical Society was organized at the Moscow Imperial Technical School (MITS), today called the Moscow Higher Technical School. It issued a *Bulletin* which, together with the *News Review* of the Technological Society, served as a basis for the *News Review in Engineering*, first published in 1915.[1]

The original charter of the Russian Technical Society affirmed the need "to ensure the continuity of education for MITS graduates with the help of common research and practical work based on faith and high morals, enabling them to exchange their knowledge, to keep abreast of science and technology, and do all they can to advance them in Russia" and "to maintain lively contacts between MITS and its former graduates, render monetary assistance to graduates and their families in need of money, and award stipends and grants for RTS students." Spearheading this development was the Polytechnical Society's first president and MITS director, Viktor K. Della-Voss, a brilliant person much admired by young engineers of his time. He was truly the soul of the society.[2]

By the early twentieth century the corps of Russian engineers had become professionally mature and well-organized. Engineering had begun to attract increasing numbers of talented young people. The Imperial Technical School, based in the Trade School of the Moscow Orphanage, obtained higher status and new civil rights after it was reorganized under the Ministry of Public Education. Upon graduating from the ITS, young engineers who did not belong to the upper classes by origin were entitled to become private honorary persons without having to pay taxes. Those who performed well as technicians or who served as managing directors of plants and factories for a ten-year period could be granted hereditary, honorary citizenship by the Minister of Public Education.

During the first two decades of this century, then, Russian engineers were increasingly well-educated, socially well-placed, and politically esteemed. Their publications focused upon not only technical but also humanitarian and cultural subjects, a sign of their increasing sense of professional self-awareness.

Nevertheless, many progressive engineers expressed dissatisfaction with the level of humanistic education in Russia. As the Russian engineer and technical philosopher Pyotr K. Engelmeier wrote in 1915:

> Many of us still have vivid memories of the time when the job of an engineer was assessed from a purely technical point of view. However, this was an erroneous opinion even then, and today the incorrectness of that narrow approach to the engineer's social function is recognized by all. History itself brings him from stuffy workshops to the fore of social activity and places him increasingly near the helm of the state. If we follow in Plato's footsteps and allow ourselves to dream of an ideal state, then we can easily conclude that just as Plato placed Philosophers at the helm of the slave state, then Engineers will inevitably have a leading role to play in the state of the present day.
>
> The engineer will have to be prepared for the leading role on a governmental level, and taking into account four crucial aspects: specifically, the technical dimension of engineering in close connection with economic and juridical concerns. However, he must also always remember his ethical function in society.[3]

Engelmeier's words are equally appropriate today; we still experience a deplorable narrowness in thinking among many engineers along with preference for rigid, technocratic policies.

Russian industries continued to expand during the early decades of the twentieth century. But a new context for the development of engineering arrived with the coming of the Russian Revolution. At first the prospects for the profession seemed to be a continuation of existing trends. In an essay devoted to the fortieth anniversary of the Polytechnical Society, its president, the MITS Professor P. K. Khudyakov wrote: "New factories and industries emerged in Russia during 1915 and 1916; they will also go on extending in the future, notwithstanding the destruction brought about by the 1917 revolution."

The All-Russian Union of Engineers was organized in March, 1917. It was renamed the All-Russia Association of Engineers (AAE) in 1918, later to be renamed again the All-Union Association of Engineers by the seventh All-Russia and first All-Union Congresses convened in 1926. The foundation of AAE was actually sealed by the Soviet of People's Commissars in a resolution, "On Measures to Raise the Level of Technical Training in the Country and Improve the Living Standards of Engineers and Technicians in the Russian Federation," published in the November 7, 1921, issue of *Izvestia* of the AUEC and signed by none other than SPC chairman, V. Ulanov (Lenin). Two years later, the organization's charter was adopted by the third Congress of Delegates. In his report, "The Unity of Labour and Science," delivered to the AAE congress in 1924, A. I. Rykov, chairman of the U.S.S.R. Soviet of People's Commissars, declared that, "Engineers have an essential role to play nowadays."

In 1925, a joint Russian-German Society for Culture and Technology was founded with A. I. Rykow and Albert Einstein as its honorary presidents. As Engelmeier wrote, "The Society's goal is to disclose and use the cultural and educational aspects of technology, to raise the level of culture and education through disseminating technical information and by upgrading cultural and technical standards. The Society seeks to popularize technical information and the latest achievements and cultural and material values. . . . The Society for Culture and Technology serves solely to achieve moral, cultural and technical purposes."[4]

Thus, in its early days the Soviet government attached great importance to expanding and improving the nation's corps of engineers. Attempts were made to define the role of technical professionals in society, not only as exponents of advanced technical ideas, but also as well-educated, culturally literate people. Such tasks were also the focus of a study group concerned with general technical issues established under the AAE branch in Moscow in 1927 by Engelmeier, who was seventy-two at the time.

Unfortunately, both the study group and the All-Union Association of Engineers survived only until 1929, when the political situation changed dramatically. The well-known policies that deformed social order in the U.S.S.R. at that time also wrecked the prestige and political standing of the engineering profession. The dreary events were mirrored in the pages of the *Engineers' Bulletin*. Its final issue commemorated Joseph Stalin's fiftieth birthday in 1929 and carried the following prophetic words: "We shall root the vestiges of sabotage, reactionary attitude and production inertness out of our environment."[5] This was preceded by a notice published in the *Bulletin* earlier that year announcing the decision to dissolve the AAE and to transfer its offices to the state-controlled All-Union Bureau of Engineers and Technicians.[6]

Why were these steps taken? The All-Union Association of Engineers was accused of being cut off from workers and technicians, of corporative and elitist trends because its members held diplomas in engineering. In addition, the engineers were accused of nothing less than aiding and abetting saboteurs and reactionaries. Thus, on the occasion of the thirteenth anniversary of the October Revolution, an engineering journal editorialized:

Mass scientific and technical societies (STS) have been set up to replace the technical societies inherited from the capitalist past and headed by the narrow, corporative AAE. Those outdated technical societies, that have ceased to exist nowadays, consisted of limited numbers of engineers in speckless clothes, isolated from the country's social and economic life and the tasks it was facing. If at the last stages of their existence, they showed interest in that life, it was solely for the purpose of scientific substantiation of its members' destructive activities. The newly established scientific and technical societies will unite on a socialist scientific basis, all groups of the working people, including workers and technicians, engineers, students and professors, serving as a true link between science and production.[7]

In this manner the scientific and technical societies were set up to replace the AAE, subsumed under trade-unions, and based on the principles of the separate trades and their narrow occupational definitions.

These developments destroyed previous efforts to build a common professional self-identity for all engineers irrespective of fields of specialization. Previously, magazines and journals of the profession had included articles on humanitarian subjects along with writings on specific technical matters. With the change in political focus, however, these publications carried militant editorials that praised the "great engineer of human souls"—"great engineer of socialist construction," "socialist architect of genius"—namely, comrade Joseph Stalin. An article entitled, "For Bolshevist Science and Technology," proclaimed, "Integrals and a cudgel, philosophy and a hammer, research laboratory and a factory, research paper and a rifle, technology and Marxism—are all links of one and the same offensive against what is left of capitalism in our country; all these are weapons of our onslaught on age-long backwardness."[8]

In their activities and publications of the 1930s, the STS and *Bulletin* concentrated on campaigns against sabotage among engineers who were now called the "senior officers of capitalism." As compared to what the STS believed to be overwhelming popular enthusiasm for the new state policy, "the treacherous nests of saboteurs" were said to stand out as "dirty spots." At an AAE congress there was a unanimous demand that the organization should be dissolved "as the last vestige of bourgeois engineers, saboteurs, and pseudo-scientists." The *Bulletin* published articles by Stalin himself in which the leader argued that the early reconstructive period "was characterized by the sabotaging disease among the most expert technicians from the ranks of old intelligentsia."[9]

Articles with titles like, "The Face and Role of Anti-Soviet Engineers," attempted to whip up fears about technical professionals secretly organizing to undermine the revolution.[10]

In 1933-1934, N. I. Bukharin's contributions to this campaign were published in *Socialist Reconstruction and Science*, a magazine dealing with the problems of scientists and engineers. "Putting production on a scientific basis and developing applied sciences is our new slogan," he wrote. Such efforts would be "headed by comrade Stalin, proclaimer of technical, scientific, and cultural progress."[11] In another article Bukharin stressed, even more pathetically: "The army of militant builders is on the march with flags flying, the sounds of the Internationale reverberating in the air, and the leader of the army is its revolutionary Field-Marshal, its teacher in his grey greatcoat, setting the pace with his will of iron—JOSEPH STALIN."[12]

Today we are well aware of where Stalin was leading the country. But at the time many people took these developments as an inspiring prospect for the large-scale remolding of society.

The so-called Coal Mine Case of 1928 was the first portent of what would eventually become a long chain of reprisals. It concerned alleged sabotage in the coal-mining industry, "exposed" in the Shakhty district of the Donetsk region. The case was launched as the start of a host of similar trials of engineers charged with sabotage in gold and platinum, textile, oil-extracting, peat and other industries. Thus the way was paved for dealing with more serious cases on a larger scale in years to come. The All-Union Association of Engineers was eventually identified as the real culprit, and its case was dubbed "trial of the Industrial Party."

The AAE had emerged following the February revolution. Its proclaimed program was "to work against manifestations of social anarchy that destroy the country's economic life such that its members cannot perform their duties, freely and conscientiously, in guiding and promoting Russian industries and technical life." The AAE was not only an organization of professionals, but also a political body uniting cultivated people with a keen feeling of self-awareness. For that reason it became a ready target for destruction by Stalin's policies, one of the first tests to see whether society was ready to follow the line he had chosen. Despite the fact that accusations against the engineers were unfounded, the first onslaught against them moved forward with great fanfare. Congratulations poured in to OGPU (State Political Department) from all over the country, along with demands for the death penalty for "counter-revolutionary bandits." "We demand that the Supreme Court of the USSR should wipe out the wreckers broken up by OGPU. Wolves cannot be chained."[13] Among other purposes, these public statements were obviously intended as a test of world public opinion.

The denunciations continued for several years. As a prominent engineering magazine explained, "One after another numerous wrecking groups in quite a few industries have been disclosed by the OGPU in the last two years. The coal-mining trial of saboteurs in Shakhty was followed by the exposure of a similar group in the railways. After sabotage on the transport, there were disclosures of wreckers' groups in the military, textile, shipbuilding, machine-building, chemical, gold, oil-extracting and other industries."[14] A united, organized guiding center was finally "uncovered" within the Supreme Council of the National Economy and the State Planning Committee, a conspiracy "connected with world capitalism," as well as "involved in the spying activities of foreign states' military headquarters." Of course, this was said to express not merely the aims of a group of counter-revolutionary engineers, but "a method of bourgeois class struggle." The "frantic ballyhoo" in the capitalist press was taken as confirmation of the "close ties" of the so-called Industrial Party (or Council of Engineers' Organizations) with ties to international organizations of former Russians and foreign capitalists. What officials of the Soviet state called the Central Committee of the Industrial Party became a convenient scapegoat, accused of organizing the 1930 economic crisis.

Thereupon followed the planned, even methodical annihilation of the corps of Russian engineers. Calls to enhance vigilance and root out the enemies of socialism brought about widespread suspicion, and not just among the engineers.

It is worth noting that during this period engineers ceased to discuss general technological problems and limited themselves to their respective narrow fields, but that did not enhance their survival. Very frequently, errors and breakdowns in technical operations were taken to be intentional acts, motivated by ideological opposition. Thus, a discussion began in the pages of the *Bulletin of Engineers and Technicians* to determine whether the forces of inertia were real or imaginary. Gradually, the question came down to one of ideology rather than of evidence; those who thought the problems were imaginary were considered to be idealists, while the supporters of the opposing view were praised as materialists in good standing.

Because they held rights and titles granted under czarist rule, a great number of Russian engineers belonged to the nobility; in the old regime there were no firm distinctions between noblemen and honorary citizens. From the viewpoint of Soviet officials during the 1930s, an engineer's nominal nobility was a sign of an obvious flaw, one that could justify moving such a person into "reforging." "Even the scum of the past, human 'throw-outs' of socialist construction, Soviet criminals, in the firm hands of NKVD (People's Commissariat of Internal Affairs), are being reformed through labour and even improve their technical qualifications at the site of the Byelomoro-Baltiisky Canal and other construction sites."[15]

Begun in 1931 as a project to link the White and Baltic Seas, the Byelomoro-Baltiisky Canal was a gigantic social experiment carried out on the bones of more than a hundred thousand people. Among its goals was that of reforming the lives and consciences of all who worked on it. This "outstanding victory of the Soviet corrective labour policy" was achieved under the guidance of the OGPU. As the Soviet Union launched this huge economic project, the former "dregs of society" were to be transformed into courageous builders of socialism. An ideal "new man" was depicted in the newspaper of the OGPU cultural and educational department at the Byelomoro-Baltiisky camp—a man with a pick and a woman wearing tarpaulin boots with a shovel in big, calloused hands. The poet Demyan Bedny was pathetically right when he referred to the Byelomoro-Baltiisky Canal in joyous terms as "the symbol of Stalin's concern." It was, after all, Stalin, the Father of Nations, who initiated the GPU labor communes and corrective labor policy in camps. It was also his idea to force prisoners to build the canal.

Engineers experienced Stalin's care and concern in other ways as well. The first special designers' offices (SDO), called "sharashki" in low colloquial, appeared during the 1930s and employed the forced labor of convicts. The aircraft designer Tupolev and general designer Sergei Korolev were among those employed in a "sharashka" when they served their penal sentences. As the design offices became widespread throughout the country, they were described enthusiastically in political speeches and the press as "the beginning."

As a result of Stalin's "concern," thousands upon thousands of lives were lost. But many people decided to focus only upon the new, limitless prospects for socialism and the rapid pace of material achievement. "Reforging" was meant to help engineers "overcome the conditioned reflexes of the capitalist era." The process of drawing sabotaging engineers to socialism went on "under high social pressure" accelerating the "thinking and nervous reactions" and reforming engineers "biologically."

Reporting on workers at the Byelomoro-Baltiisky Canal, one official commented, "When at liberty, they used to slow down the pace of their work deliberately. There could be over-subtle, throw dust in one's eyes, quote someone as an authority, while here they are convicts, and there is no throwing dust in security men's eyes. The Party and government decision about the terms of construction is irrevocable, and all kinds of false rumors are out of place." True, the conditions were hardly fit for work; laborers were placed in confinement, "at the world's end," "with the shortage of metal and cement, and without a single skilled worker" and with no "complete technical design." The job did not leave any time for thinking. "The plan, the inexorable working plan, turns gradually into the supreme law that must equally be obeyed by security men, engineers, thieves, bandits, and prostitutes."[16]

Such was the "paradise at the world's end, under the skies of Karelia and GPU guardianship" where the engineers experienced "a great joy of life." Those were the engineers, the "general technical headquarters of Russian capitalism," presumed saboteurs, who found themselves in the ranks of the enemies of the Soviet government. They were arrested by the OGPU, pleaded guilty, and expressed readiness "to begin working for Soviet power," i.e., to build the canal.[17]

But of what crimes were the prisoners guilty? The report makes it clear that one of the despised offenses was an improper use of language.

An indication of the "haughtiness of a caste of engineers" was that they avoided using abbreviated "Soviet words." According to one official, "Whenever possible, engineers tried to avoid using them. More often, they said 'institutions of higher learning,' and not IHL [VUZ in Russian], 'five-year construction plan,' instead of 'pyatiletka' [five-year period in abbreviation], 'workers faculty,' instead of 'rabfak'. . . . This [way of talking] meant just one step toward sabotage, which they took with great ease." The prisoners' experience at the Byelomor Project "was not wasted on them," and they "got rid of the shortcomings implanted in them by the corrupted social milieu in which they had mostly lived and worked. Engaged in socialist labour, they were able to develop their merits to the full." "Their intricate and painful subtleties were forgotten: their new position made them adapt to the bolsheviks."[18]

It is now clear what the "direct road" offered cultivated Russian engineers: a slashing attack against the AAE, trials aimed at rooting out the Industrial Party, and "reforging" at the Beylomor construction project. In this way the engineers' corps of the young Soviet Republic, its elite, was smashed and reduced to almost nothing. In the process, Russia's technical capabilities were leveled and engineers' professional pride and feelings of self-awareness trampled upon. Because this effort was carried out so systematically, later administrative attempts to repair the damage could do little to enhance the prestige of the intellectual work of engineering.[19]

The broader lessons from this sad chapter in the history of science and technology in the Soviet Union can now be summarized. They are essential not only for today's Russia but for the world as a whole.

1. Genuine scientific and technical progress is inconceivable without freedom and democracy in all spheres of social life.

2. Hard and fast planning from above cannot benefit society. It leads to anarchy of production and places consumers under producers' *diktat*, because it is impossible to foresee all the details from above. A self-regulating economy is needed, with a reasonable combination of planning and the market.

3. Centralized manipulation of huge resources, transferring them by orders and decrees (without outside social appraisal of experts and regardless of public opinion) is not advantageous, but detrimental to nature, society, and human beings. This is demonstrated by the consequences of attempts to change the flow of rivers and crash programs for building nuclear power stations.

4. There is no way to make people happy from above against their will and without their agreement and active participation. Employing forced labor in general, and engineers' labor in particular, is not only

immoral, but also inefficient.

5. Genuine progress in science and technology—demonstrated less in specific breakthroughs than in gradual improvements in the standard of people's everyday lives—cannot be achieved without a high level of humanitarian culture in society.

Foremost in the development of today's engineering profession is an education in cultural contexts of technical work. The lack of a vital culture of engineering of this sort is not as inconsequential as it might seem, because the diploma of an engineer enables one to gain all kinds of leading posts in society. In such positions, narrowly trained engineers apply their narrow technical approach to human beings, thus becoming "engineers of human souls" in the worst sense of the word, as up-and-coming bureaucrats. This is why so many ecological, sociopsychological, and other problems are closely associated with engineering practice nowadays. This is the cause of dangerous technological distortions, unlimited overloading of technology, and of inhuman attitudes toward people. It is also the root cause of the technological accidents and ecological disasters so frequently highlighted in the press in recent years.

Contemporary engineers must deal with both human and machine systems. For that reason they have to be well-read in the humanities as part of their professional training. In an article, "Do We Need Philosophy of Technology?" written in the fateful year 1929, P. K. Engelmeier asked, "What are the demands placed on engineers?" His answer: "broader horizons, deeper speculation, philosophical thinking, better economic, sociological and historical knowledge, and at the same time, engineers must remain engineers in their ambitions and mentality."[20] His words are as vital for us now as they were decades ago. As Engelmeier observed two years before the October Revolution, "The time is past, when engineers had to work solely inside workshops, and when their technical knowledge was all that was required of them. The expanding industries demand that their leaders and organizers should be not only technicians, but also lawyers, economists, and sociologists."[21]

Institute of Philosophy, Academy of Sciences, Moscow

NOTES

[*Editor's note*: The footnotes to Vitaly Gorokhov's paper are sometimes incomplete. I present them in the form he used in his original paper, one that I rewrote to achieve greater clarity in its English prose. Checking these references with the author proved impossible during the turmoil that enveloped the Soviet Union during 1990-1991.]

1. When the All-Russian Association of Engineers was founded, this journal was issued under its aegis.
2. "40 Years of the Polytechnical Society Attached to the Moscow Technical School" (1917), p. 7.
3. Pyotr K. Engelmeier, *News Review in Engineering*, no. 3 (1915), p. 294.
4. P. K. Engelmeier, "Culture and Technology," *Technical and Economic News Review* 5:7 (1925): 515.
5. *Engineers' Bulletin*, nos. 11-12 (1929), pp. 382-383.
6. At this point the *Engineers' Bulletin* was also moved to a bureau of the Soviet state to be published under the new title, *Bulletin of Engineers and Technicians*.
7. *Bulletin of Engineers and Technicians*, nos. 11-12 (1930), pp. 362-363.
8. *Bulletin of Engineers and Technicians*, no. 4 (1931), p. 142.
9. *Bulletin of Engineers and Technicians*, no. 11 (1934), p. 495.
10. *Engineers and Their Work*, no. 8 (1928), pp. 361-362.
11. N. I. Bukharin, "Technical Reconstruction and Current Problems of Research," *Socialist Reconstruction and Science*, SORENA, no. 3 (1933), p. 30.
12. N. I. Bukharin, "Science and People: Heroic Symphony," SORENA, no. 2, (1934), p. 21; section III, "Builders."
13. "We Are on the Alert," *Peasant Newspaper* (1931), pp. 22.
14. *Engineers and Their Work*, no. 21 (1930), p. 662.
15. Y. Rykachev, "Engineers at Byelomor Project," *Ogonyok Library*, no. 16/787 (1934), p. 8.
16. *Ibid.*, pp. 6, 3, and 13.
17. *Ibid.*, pp. 43, 38, and 20.
18. *Ibid.*, pp. 32 and 40.
19. Science cannot develop in isolation, spurred by administrative pressure and command. The same is true of the development of industry and society as a whole. In the light of today's pain and hope, it is worth remembering a declaration issued by the 1904 constituent assembly of the All-Russia Union of Engineers and Technicians: "Russian industries can develop successfully only on the basis of public and personal initiative over a large scale. The prerequisites of such development include: complete inviolability of person, freedom of assembly and association, of speech and the press. Only a reliable guarantee of those rights can facilitate better and broader public education, without which higher labour productivity is inconceivable." Declaration of the All-Russia Union of Engineers and Technicians (1906), p. 1.
20. P. K. Engelmeier, "Do We Need a Philosophy of Technology?," *Engineers and Their Work*, no. 2 (1929), p. 36.
21. P. K. Engelmeier, "Defending General Ideas in Technology," *News Review in Engineering*, no. 3 (1915).

TOM ROCKMORE

HEIDEGGER ON TECHNOLOGY AND DEMOCRACY

One of the most interesting and troubling facts about Martin Heidegger, a realization brought forcefully to our attention by the publication of Victor Farias's study of Heidegger and Nazism, is the fact that, after all the qualifications and interpretation, this important German thinker was deeply and irrevocably committed to one of the most anti-democratic movements of this or any other historical moment.[1]
The problems Farias expounds had been identified previously by others, including the gifted but dogmatic Marxist Georg Lukács. One is tempted to dismiss Lukács's opinion because of his dogmatic attitude, but this would be a mistake. On examination, we see that his early, in fact nearly prescient reading of the problem of Heidegger's relation to fascism looks better now than it did when Lukács wrote.[2] Unlike some others, who at this late date often still hesitate to draw clear conclusions about Heidegger, Lukács was never constrained by such intellectual niceties.[3] One need not accept his harsh verdict on Heidegger as the Ash Wednesday of subjective idealism.[4] Lukács is also incorrect in labelling Heidegger as merely pro-fascist in view of the latter's public and apparently sincere commitment to the essence of National Socialism, as witness his famous reaffirmation of the significance of the movement as late as 1953.[5] This situation invites sober reflection on the extent to which claims for the social relevance of philosophic reason, claims that run throughout the philosophic tradition from Socrates to the present, are realized in the case of this brilliant but sociopathic thinker. In particular, we need to inquire if Heidegger's somber pessimism is justified with respect to the phenomenon of technology.[6] And we need to join Lukács in raising an equally important issue, namely whether Heidegger, under the cover of the question of the meaning of Being, has not transformed, in mythological fashion, the human condition in advanced capitalism into a universal and general human condition.[7]
A caveat is in order. In the wake of our knowledge of Heidegger's relation to National Socialism, it is not useful, nor is it my intention, to besmirch his reputation in any way. In any case, no critic of his position could do so more or more effectively than he himself has already done. Rather, we need now to rethink our attitude towards his thought without any preconditions in order to determine if, as some believe, the thinker and his thought are independent entities, or if, as others hold, the two are inextricably intertwined. Obviously this is a

complicated task, one that will require many years of careful discussion among philosophers. It would be as irresponsible as Heidegger himself appears to have been for us to prejudge the case without providing an adequate hearing. But it would be equally irresponsible to fail to raise questions of this kind. Indeed, Heidegger seems to invite this type of examination of his thought on specific grounds, including the general insistence on *Dasein* as inevitably situated and the specific claim, in the course of the *Spiegel-Gespräch*, that his own thought must be judged in terms of his political engagements.

The aim of this paper is to begin to undertake this task as concerns Heidegger's view of technology. It is this same view of technology which has recently been described by a revisionary critic as the most powerful part of Heidegger's corpus where everything comes together.[8] We need to decide whether Heidegger's reading of technology can stand as a permanent contribution by an important thinker. We need to ask ourselves whether this reading is adequate or fundamentally flawed.

Four points will help to situate Heidegger's notion of technology within its conceptual and historical context. First, the discussion of technology occurs in Heidegger's thought after the end of the war. This discussion, which follows Germany's defeat in the Second World War, exhibits a deep pessimism, particularly in terms of the relation of human beings to technology. But we will need to determine if the pessimism which is almost a constitutive part of Heidegger's thought in both its first and second periods is based on anything more than his subjective reflection of the existential situation he described in *Being and Time*. Thus, we will need to evaluate the reasons Heidegger advances to justify his pessimistic perspective.

Second, explicit attention to the concept of technology only engages Heidegger's attention after the statement of his initial position, as it evolved after the publication of *Being and Time*. The transition to Heidegger's later thought occurs through the so-called turning (*Kehre*), in which he maintained his concern with the problem of the meaning of Being, the so-called *Seinfrage*, even as he modified the approach to it. Almost everything about the turning, including its exact nature, location in Heidegger's writings, as well as its philosophic significance, is controversial.[9] Now it seems normal to expect that Heidegger modified his position in order to improve it. But since he also maintains that a true philosophy does not progress, the idea of the *Kehre* is problematic.[10] Either there is no change in his thought, in which case there is also no improvement, and the turning is only apparent; or there is a turning which improves the position, and Heidegger is inconsistent.

Third, there is a tension, perhaps even a basic contradiction with respect to Heidegger's own notion of ontology. His question of the meaning of being, which he labels as ontology, is in fact an epistemological query, since his understanding of ontology is not the same as, say, the kinds of concerns which Aristotle raised in the *Metaphysics*. Suffice it to say that Heidegger presents a modern view of metaphysics similar to that of Descartes, Kant, and Hegel, for whom metaphysics has become epistemology as opposed to ontology. But epistemology requires a concept of subjectivity, in Heidegger's terminology, *Dasein*. Hence, there is a tension in Heidegger's reliance on *Dasein* to pursue the study of ontology as he comprehends it and his desire to transcend a concept of the subject in order to let Being unveil itself directly without the mediation of *Dasein*.

Fourth, the discussion on technology provides a kind of litmus test for Heidegger's understanding of his relation to Nazism. He turned to technology after a period as Rector of the University of Freiburg. Heidegger's explanation of his relation to National Socialism after he resigned his post as rector is at least controversial. His conception of technology arises in the wake of his famous profession of faith, not in Nazism as it occurred, but in the idea of Nazism, what, in its revised form, he called "the inner truth and greatness of the movement."[11]

This statement provides an important indication of the presence, even in Heidegger's thought after the *Kehre*, of a continuing concern with the distinction between authenticity and inauthenticity, here applied to Nazism. In order to reject other interpretations of National Socialism as failing to grasp its essence, Heidegger commits himself to the capacity to differentiate the non-essential from the essential, in short to provide an authentic interpretation of National Socialism. This statement further explains the later failure, even in the famous *Spiegel* interview, to distance himself from the holocaust. For to mention the holocaust, to acknowledge that National Socialism was not incidentally but fundamentally linked to one of the most enormous crimes in the history of the world, would have required Heidegger to acknowledge that the possibility he saw in National Socialism, namely as concerns the defense of the German university and the coming to fruition of the German people, was an enormous and deep error. Whether Heidegger was aware of this fact and desired to distance himself from National Socialism as it occurred is an important question. Some have suggested that the discussion of technology belongs to a phase when Heidegger was trying to distance himself from Nazism.[12] But if this was Heidegger's intention, he certainly did not succeed, since the discussion of technology is one of the less pellucid texts by this difficult thinker.[13] In fact, nothing in this text suggests that in writing this essay Heidegger was trying to take stock in any straightforward way of what had occurred and what his own role had been.

For historical reasons, Heidegger's view of technology is forever linked with his turn towards National Socialism. The famous *Spiegel-Gespräch,* intentionally published only after Heidegger's death, and which was perhaps intended to provide Heidegger's own interpretation of his legacy, to function as a kind of testament, begins with a series of questions about Heidegger's relation to Nazism. The interview then turns to a question about Heidegger's statement, made in 1935, about "the inner truth and greatness" of National Socialism. In 1953 the lectures in which he made this notorious statement were published as the *Introduction to Metaphysics.* In this acclamation for the National Socialist movement, the words, "namely with the meeting of planetarily limited technology and modern man," are placed in parentheses. In the interview, Heidegger defended himself against the accusation that he had only later added the words in parentheses, which, he claimed, were present in his original manuscript.

The interview then turns to Heidegger's view of technology. Several of his statements are worth noting here. He admits that he is unclear which political system fits modern technology and he states that he is not convinced that it is democracy. He states that human being is not in control of technology. And he asserts that at present philosophy cannot bring about a change and only a God can save us. He goes on to say that traditional thought, which he distinguishes from his own view under the heading of the other thought, is unable to comprehend the technological age. For Heidegger, the task of this other form of thought is to aid mankind in bringing about a satisfactory relationship to the essence of technology.

These comments on technology and National Socialism bring together two elements in Heidegger's thought that are not entirely clear, and whose relation is, to say the least, troubling. In the *Spiegel* interview Heidegger suggests that his statement about inner greatness and the movement has been misunderstood and that it is actually linked to his own desire to do battle against modern technology. He argues that we cannot even think technology with ordinary philosophical thought, although such thinking is possible from the perspective of his so-called other thought which lies beyond metaphysical thinking. And he expresses doubts that democracy is relevant at this stage in the development of mankind.

For my purposes here, I will confine my comments to an initial discussion of Heidegger's view of the nature, or to use his term, the essence of technology. As an aid in this task, it will be useful to recall the development of the theme of technology in Heidegger's position. Technology is not an important theme in *Being and Time.* So far as I know, the only explicit mention of technology is in a single sentence in paragraph 69, which is a lengthy discussion of "The Temporality of Being-in-the-World and the Problem of the Transcendence of the World." In part b of this paragraph, called "The Temporal Meaning of the Way in which Circumspective Concern becomes Modified into the Theoretical Discovery of the Present-at-Hand Within-the-World," Heidegger writes: "Reading off the measurements which result from an experiment often requires a complicated 'technical' set-up for the experimental design."[14] But the basic conceptual framework within which Heidegger's analysis of technology will shortly emerge is already in place as early as this work in the famous discussion of equipment, where from a clearly pragmatic angle of vision he draws the well known distinction between readyness-to-hand [*Zuhandenheit*] and presence-to-hand [*Vorhandenheit*]. He then says, in a reference to what nature is in itself:

> Here, however, "Nature" is not to be understood as that which is just present-at-hand, nor as the *power of Nature*. The wood is a forest of timber, the mountain a quarry of rock, the river is water-power, the wind is wind "in the sails." As the "environment" is discovered, the "Nature" thus discovered is encountered too. If its kind of Being as ready-to-hand is disregarded, this "Nature" itself can be discovered and defined simply in its pure presence-to-hand. But when this happens, the Nature which "stirs and strives," which assails us and enthralls us as landscape, remains hidden. The botanists' plants are not the flowers of the hedgerow, the "source" which the geographer establishes for a river is not the "springhead in the dale."[15]

This passage already exhibits a claim characteristic of Heidegger's later writing, including his discussion of technology. In his account, there is an essence that surpasses, or lies beyond, his familiar distinction between readiness to hand and presence to hand. One instance of this essence is that of Being in general. Another instance is the essence of technology. Heidegger develops this idea further in his *Introduction to Metaphysics* (1935), where he equates Russia and America as two instances of the same domination of man by technology.[16] The important point is that the essence of technology, a still unknown essence, influences us greatly. This view is quickly expanded further in a series of later publications. For instance, in his two-volume study of Nietzsche, Heidegger discusses in passing the technological style of modern science and emphasizes the calculative reason of technology, a point which clearly goes back to Max Weber. And in the important essay entitled "The Age of the World Picture" (1938), he suggests that mechanization is the most visible extension of the essence of modern technology, which he describes as identical to modern metaphysics.[17]

The important point here is the link Heidegger tries to establish between technology and metaphysics. We know that his whole position is dedicated to the renewal of the problem of ontology, and we further know that for Heidegger this requires the overcoming of metaphysics. Since he here indicates, in reflecting on the nature of modernity, that machine technology is in some undefined sense identical to modern metaphysics, we can infer that for Heidegger the problem of technology is not merely an eccentric preoccupation but is instead central to his thought.

Heidegger develops his view of technology further in a short passage in the "Letter on Humanism" (1947). He comments on Marx's theory of alienation, which he describes as superior to other views because it attains an essential dimension of history. He then writes:

> For such dialogue it is certainly also necessary to free oneself from naive notions about materialism, as well as from the cheap refutations that are supposed to counter it. The essence of materialism does not consist in the assertion that everything is simply matter but rather in a metaphysical determination according to which every being appears as the material of labor. The modern metaphysical essence of labor is anticipated in Hegel's *Phenomenology of Spirit* as the self-establishing process of unconditioned production, which is the objectification of the actual through man experienced as subjectivity. The essence of materialism is concealed in the essence of technology, about which much has been written but little has been thought. Technology is in its essence a destiny within the history of Being and of the truth of Being, a truth that lies in oblivion. For technology does not go back to the techne of the Greeks in name only but derives historically and essentially from *techne* as a mode of *aletheuein*, a mode, that is, of rendering beings manifest. As a form of truth technology is grounded in the history of metaphysics, which is itself a distinctive and up to now the only perceptible phase of the history of Being.[18]

I have quoted this passage at length because it suggests important ideas which will reappear in his later discussions of technology. It also gives clear indications of the limits of his overall view. Heidegger presents his concept of technology in the context of remarks on the absence of a grasp of history in such other forms of phenomenology as those proposed by Husserl and Sartre. He prefers what he calls Marx's metaphysical view, one derived from Hegel. The superiority of Marx's position, Heidegger claims, lies in its superior understanding of history. This suggests that Heidegger accepts an adequate comprehension of history as a criterion for the evaluation of his own theory. But he rejects the efforts of Marx and Hegel to grasp history because they rely on the category of labor, which he describes as a metaphysical principle. Since he wants to surpass metaphysics, the theory of history he accepts must be non-metaphysical.

Heidegger follows a widespread presupposition in the secondary literature, strongly stressed by representatives of Marxism, of a difference in kind between idealism and materialism. Rejecting what he believes to be a metaphysical approach because it is metaphysical, Heidegger advances what he believes to be a so-called third way that will surpass the split between Hegel and Marx, between idealism and materialism.

Marx regarded modern technology as something that arises within capitalism as a function of the process of capital accumulation. For Heidegger, technology is a metaphysical occurrence, whose essence has often been discussed but not often thought. In order to grasp the essence of metaphysics, he believes we need to connect technology to truth as found in the ancient Greek view of *techne*. He maintains that there is not merely an etymological link between *techne* and technology, but also an essential link; both are elements in the history of metaphysics, which itself belongs to the history of Being. Thus Heidegger argues that technology is an occurrence in the history of metaphysics and of Being which we need to comprehend from a perspective which attends to Being as the central theme, but from an angle of vision that surpasses metaphysics.

Now we must comprehend his understanding of what is at stake. He explains his view in a remark about why Communism and Americanism are respectively more than a *Weltanschauung* or a life-style. According to Heidegger there is a relation between metaphysics and the destiny of Europe; for Europe is falling behind in what he obscurely refers to as the dawning of world destiny. Metaphysical thought is incapable of getting a hold on this destiny. His claim can be summarized as follows: there is such a thing as the destiny of the world, which is somehow related to Being; Europe has failed to understand the destiny of the world because it approaches the problem through metaphysical thinking. Heidegger offers an allegedly post-metaphysical thinking, said to be useful in enabling Europe to "catch up" by grasping an object of knowledge which cannot be known through metaphysical thought.

Clearly, Heidegger is making a radical claim for the relevance of his own form of reason. Elsewhere he suggests that in the true sense thought exceeds both theory and practice through its inconsequential nature.[19] This humble declaration only partly masks the ambitious nature of his claim for thinking. Since the phenomenon in question is the essence of technology, we can infer that he holds that the fate of Europe, in fact even world destiny, depends on the turn to a new, non-metaphysical form of thought alone capable of comprehending technology. For Heidegger, then, the idea of technology is not a mere object of bewilderment, a philosophical conundrum, of interest to the masters of abstract thought. Rather, philosophers, whether idealist, materialist, or Christian are simply incapable of understanding technology. The understanding of technology is clearly fraught with a fateful consequence, since, in Heidegger's view what is at stake is nothing less than the capacity of Europe to keep pace within the fate of the world.

The main source of Heidegger's mature view of technology appears in a single essay entitled, "The Question Concerning Technology" (1954). Confining the discussion in this way is limiting, I am aware; a full discussion of all the material on Heidegger's view of technology would need to consider a variety of other writings. My present aim is to come to grips with Heidegger's view of technology, in effect to think with him on his own level in order to determine the outlines of his view of technology and to evaluate it. In contrast to Heidegger, who did not believe in criticizing another thinker, I see no reason to shrink from an evaluation of his (or any other) theory, no reason to eschew an evaluation of his view of technology.

To a degree unusual even in Heidegger's writing, the discussion in "The Question Concerning Technology" is murky. If, as Pierre Aubenque has claimed, Heidegger actually wanted to criticize National Socialism in this essay, it seems reasonable that he would have made a greater effort to be understood. The discussion is repetitive and obscure, producing just the kind of hermeneutical issues which occupy scholars concerned (so to speak) more with the trees than with the forest, who in their attention to such scholarly minutiae become unlikely to raise larger questions about the meaning of his position.

Heidegger is constantly attentive to language, seeking to unveil what he regards as original meanings that have been covered up over time. From that viewpoint it is interesting that Heidegger entitles his discussion "Die Frage nach der Technik" and does not utilize the German term which literally means "technology." The word *Technologie*, which Heidegger does not employ here, derives from *Technik* and means roughly the knowledge of technique applied to a mode of production. The English term "technology" in the title of the essay translates *Technik*, but its closest equivalent in English is "technique," probably not "technology."

There is a difference in meaning here which is obscured by the translation. The English word means something like "the application of scientific knowledge to practical purposes in a particular field," or "a technical method of achieving a practical purpose." The German word means something like "the art of reaching a determined goal with the most purposeful and economical means," for instance the totality of means by which nature is rendered useful to man through knowledge and application of its laws. It includes the totality of tricks, rules, and procedures in a given domain, as well as forms of production, capacities, and aptitudes. Just as in German the term *Wissenschaft* is wider than the English "science," so there is an important difference between the domain covered by *Technik* and that implied by "technology." Whereas in English the field in view is something like applied science, in German it includes a wide range of fields such as construction work and even the theatre. In short, "technology" is not an exact translation of *Technik*. This helps to account for a certain unexpected quality in Heidegger's account of technology which tries to elucidate a familiar phenomenon, namely technology, under the unusual heading of the term "technique."

Heidegger's discussion proceeds in his familiar apodictic style, mainly the absence of any recognizable form of argument. Since I think there is a considerable likelihood of going astray if we remain too close to the text, I shall try to restate what I regard as the central part of his view of technology in terminology more neutral than Heidegger's own language. In *Being and Time*, Heidegger posed what he called the question of the meaning of Being. In much the same way, the essay on technology raises the question of the meaning of technology, or more precisely the question of the essence of technology, where "essence" is understood in what he claims to be a non-metaphysical sense. What he calls "essence" is in effect the essence of the conception of essence; it is something which endures, holds sway, administers itself, develops, and decays. In this connection, he calls attention to the etymological relation of the term essence [*Wesen*] to the verb *währen*.

But there is no easy way to justify or even to comprehend one strange statement: "Only what is granted endures. That which endures primally out of the earliest beginning is what grants."[20] This statement is strange since it seems to acknowledge that the essence is in some unacknowledged sense dependent on something beyond it. So in many religions, it would make sense to say that grass is the way it is because God made it, or that human being is made in the image of God. Heidegger seems to be suggesting that as part of his non-metaphysical understanding of the concept of essence, the essence is not to be understood in its own terms. To understand an essence we need to appeal to something beyond the essence, so that we can differentiate what is granted and, to employ consistent terminology, a grantor. The argument seems to rest on the relation of *Wesen* with *währen*, and of the latter verb with such further forms as *gewähren*, which, among other things, means "to grant." Further allusions to Goethe's usage of *fortgewähren* for *fortwähren*, although perhaps an interesting linquistic observation, cannot yield a conclusion in relation to the nature of essence. It is unclear why an essence needs or even could be granted, as if there were an initial agent which brought into being or maintained essences, such as Plato's demiurge. Heidegger seems to have something like this in mind in his concept of essence, but he neither clarifies nor supports this claim.

Just about the only unambiguous statement in the essay is Heidegger's passing comment that the essence of technology is ambiguous. The flow of his discussion indicates his concern to reveal the essence of technology and to differentiate technology from its essence. He notes that current concepts of technology are instrumental and anthropological, noting that technology is a means to an end but no mere means. Comments on the relation of instrumentality to causality lead to a further comment on revealing, which Heidegger relates to the concept of truth. He then asserts that technology is a way of revealing. Man does not control the unconcealment since modern technology is no human doing. A meditation on enframing—the horizon in which modern technology comes about—as the essence of the matter leads to further comments on destiny, danger, and the saving power[21]

In this way, Heidegger insists on a profound difference between the utilitarian and essential aspects of technology. His basic claim is that technology is a way of revealing. This suggests that there is something beyond technology which is revealed by it. He also claims that the unconcealment is not under human control, that it is somehow associated with destiny, and that it is further linked to both danger and salvation.

The distinction between the utilitarian and essential aspects of technology at first glance appears relatively straightforward, even if Heidegger's statement of it is not pellucid. In more ordinary terminology, he appears to be pointing to the difference between utility and truth, which some pragmatists, notably James, are accused of conflating, and which Heidegger tried to keep apart in his differentiating of readiness-to-hand from presence-to-hand in *Being and Time*.

His further remarks about the essence of technology are less obvious, in fact often extremely difficult to construe. They can be reconstructed as follows: "Technology" derives from *techne*, which means "bringing forth," or *poiesis*, in art, or the fine arts, both in ancient Greece and, apparently by analogy, in technology. Now revealing entails that there is something concealed, as well as a means of unconcealment, in this case in technology. In this respect, the significant claim is that human being does not control the unconcealment, by which Heidegger understands the process in which the real shows itself in technology. For Heidegger, human being is confronted with a double incapacity: the presence of a realm of reality, perhaps Being as such, which is independent of us; and a process in which that which is independent can show itself at any time. Heidegger argues for this position by pointing to the fact that a forester is variously controlled by profit-making in the lumber industry, the possibilities of cellulose, public opinion, etc. His point seems to be that the forester is passive with respect to the technology he employs. But this argument seems insufficient to justify an assertion that there is something like a realm of reality that is in fact revealed in technology; nor does it justify the inference that the revealing of technology itself is beyond human control. At best it shows only that it might be beyond the control of a given individual, such as the forester in the example.

It is worth noting that Heidegger locates various aspects of the technological process within social life. But at this crucial point in his discussion he fails to offer a theory of the social world. He needs to do this, however, if he hopes to understand the place of technology within human affairs. This step is also needed if he is to make good his claims that there is a further reality and that technology is beyond human control.

This lack is not a mere lacuna, an oversight; rather, it is a fundamental flaw. In defense of Heidegger, one might point out that he wants to go beyond the usual anthropological perspective in order to comprehend something deeper than or beyond the human dimension. But whether or not Heidegger can establish a transcendent dimension of technology, he still needs to locate technology in the social arena in which it arises and in which it functions. His manifest failure to do so is tantamount to a failure to analyze that aspect of the technological phenomenon which is closest to us and most familiar.

The central claim in Heidegger's view of technology is that its nonmetaphysical essence lies in enframing [*Gestell*]. I will not reproduce the tortuous etymologies through which Heidegger suggests the meaning of this term. In any case, an etymological approach is insufficient to do the work Heidegger intends for it. There is no reason to believe that there are correct meanings which have been covered up in the course of time and which can later be uncovered in order to reveal the way things are in some deeper sense. Language is a matter of convention. What the Greeks meant by the use of a given term, which survives in later language, is only what they meant; it is not for that reason the true meaning of the later word. Earlier meanings are not true since language itself is not true; it is only an instrument with which to point, to refer, and to communicate.

Heidegger's concept of enframing is linked to his idea of standing reserve [*Bestand*, from *besthen*]. If we again recall his distinction in *Being and Time* between readyness-to-hand and presence-to-hand, the term "standing reserve" seems to mean something like the way things are disposed or ordered so that they can be utilized for various purposes. Heidegger suggests as much in his example of the plane on the runway ready for takeoff. But consistent with the concern in his later thought to surpass the anthropological approach, he does not want to identify purpose with human purpose. He does not argue for this view, but merely states it. One example is his remark that as a revealing, modern technology is no merely human doing. Another illustration of this quality in his writing is his sibylline description of enframing as "that challenging claim which gathers man thither to order the self-revealing as standing-reserve."[22]

The latter claim is especially significant, because Heidegger now supplements his earlier insistence on a trans-human reality and on technology as a mode of revealing with a new stress on the active role of that reality with respect to human being. An analogy would be Hegel's view of spirit which is said to realize itself in and through human actions. For Hegel, the subject of history is finally more than human. For Heidegger, similarly, what he refers to variously as reality or Being shows itself altogether independent of human being. But since Heidegger does not offer reasons for accepting a trans-human reality, it is rather difficult to accept that it is active with respect to human being.

The inability to sustain the claim that there is a trans-human reality which allegedly acts through technology gravely compromises Heidegger's assertions about such aspects of enframing as destiny, danger, and saving power. His notion of destiny [*Geschick*] is not well articulated in his essay, "The Question Concerning Technology." The concept reaches back in his thought at least as far as *Being and Time*. In that work he takes up the discussion of the "Basic Constitution of Historicality," and differentiates fate [*Schicksal*] from destiny [*Geschick*]. Heidegger describes destiny from the perspective of the resolute future-directedness he stresses there as "the historicizing in Being-with-others."[23] But it is notable that in *Being and Time* there is neither a substantial discussion of such phenomena as danger and saving power, nor an association of either concept with his concept of destiny.

In the essay on technology, danger and saving power are associated with destiny, but neither the ideas themselves nor the purported relation is clearly elucidated. According to Heidegger, the destiny of revealing is not a particular danger, but danger as such.[24] But the nature of this general danger is at least unclear. Perhaps Heidegger has in mind death, which in *Being and Time* was described as the ultimate limit of life. If this is the case, it needs to be stated. And it further needs to be shown why revealing, or the destiny of revealing, is in general a danger, or dangerous. Conversely, Heidegger provides no ground for the association of danger with saving power other than a quotation from Hölderlin. But this is scarcely a sufficient reason.

In the present context, danger and saving power are secondary aspects, that is they seem to depend upon destiny, more precisely upon the destiny of revealing. This leaves the question of the role of destiny in the phenomenon of technology unresolved. Looking back to *Being and Time* is helpful since the earlier discussion of historicality [*Geschehen*] led to the idea of a dimension of temporal happening which stands outside of resoluteness, even resolute future-directedness. Perhaps by analogy with Greek thought, Heidegger means to associate this with what the ancients called *moira*. But this is at most a piece in the puzzle. Once again Heidegger seems to be relying on associations between terms. So in German but not in English the terms "destiny" [*Geshick*] and "fate" [*Schicksal*] have an etymological relation to the verb "to send" [*schicken*]. From this association, we can jump to the conclusion that that which is revealed through destiny is sent and history is what is sent, or rather the process of sending. As he says: "We shall call that sending-that-gathers which first starts man upon a way of revealing, *destining*. It is from out of this destining that the essence of all history is determined."[25] If this reconstruction of his chain of reasoning is correct, what we have here is a change in his position. In *Being and Time*, Heidegger grounds historicality in time. Here he seems to be suggesting that the essence of all history (*Geschichte*) is based in a kind of sending, which leads to revealing.

This proposed reconstruction can be read in two very different ways. Heidegger seems to have in mind something like the Greek concept of actions beyond man's control, of that which just happens. This kind of claim is certainly admissible. Few writers care to argue that everything without limit is or could even conceivably be under the control of human being. For instance, in a famous passage Marx writes that men make their own history, but they do not do it as they please or under conditions they have chosen.[26] But if something is sent, then there must be a sender, which leads back by this complicated route to the transpersonal agent which Heidegger regards as operative in the historical sequence. This is what manifests itself through technology. This is a recurring claim, evident in his assertion that enframing acts upon man[27] and in similar assertions in his later writings.[28] But as far as I can determine, this claim is never argued but always merely asserted.

One of the tests of any philosophy which aspires to more than specialized competence in a restricted domain is its ability to comprehend the world in which we live. One of the important aspects of the modern world is the rise of modern technology. During the last hundred years or so the elaboration of applied science has enabled the application of new techniques to attain various goals both desirable and undesirable, but practically feasible within the existing social and political framework. Technology, then, is a means to an end which human beings determine and toward which, through this means, they strive.

If this undertanding of the nature of technology is true, then Heidegger's view of the matter falls short of the mark in at least three ways. First, in his return to the Greek concept of *techne*, or art, Heidegger blurs beyond recognition the essential distinction between the concepts of art, technology, and technique. In his essay, he clearly has in mind technology in the English sense of the term, as his references to hydroelectric plants on the Rhine and aircraft clearly show. It is true that there is a poetic moment in technology, and that technology often requires using one or more techniques. To take a humble example, the technique of welding is applied in spotwelding, which is a form of technology utilized in various fields, such as plumbing. But despite the etymological associations upon which Heidegger plays, there is no reason to conflate art, technology, and technique.

It is helpful, indeed vital, to maintain the separation between a technique and a technology. Technology, like art, makes use of techniques, which are neutral with respect to both art and technology. If by "technology" we choose to mean something as general as the "practical implementation of intelligence," which in turn means that "the implementations are not wholly ends in themselves" but are "embodied in implements, artifacts, or sometimes in social organization,"[29] then we must keep apart the technology in general and its technical building blocks, or techniques.

If we take as our thesis the identification of art and technology, then it would be justified to argue that technology is no mere human invention but refers beyond itself. But since there is nothing other than a slim etymological link to point to a realm which technology reveals, we have only hortatory grounds to accept this conclusion. It may be appropriate to argue that genuine art is characterized by reference to a transcendent level of reality. But it is only on the basis of a deep conflation of art and technology that the latter can be taken for the former. This error follows directly from Heidegger's own terminology. For it is only by discussing technology under the heading of technique that he can discover a common term which links art and technology. To do so, however, is to mistake the nature of technology; for technology is not art, although some technology may have artistic results. Art is not technology, even if many forms of art routinely employ forms of technology.

In his attention to technology as the prolongation of metaphysics by other means, Heidegger fails to understand the way in which technology is embedded in modern social and historical contexts. On the one hand, his approach to technology through the problem of metaphysics sees technology not as an end in itself but as a means to a further end, in this case as part of his effort to destroy the history of metaphysics. But the terms "metaphysics" and "ontology" are not univocal. They have been understood in different ways in the history of the philosophical tradition. Aristotle, who is usually regarded as the father of metaphysics, did not use that term, which was only later applied to his writings. But he did speak of the science of being as being. Now it is clear why Heidegger draws a connection between metaphysics and technology. For metaphysics is connected to representational thinking, to what he elsewhere calls a world picture, in short to a two worlds ontology which begins in Plato. Technology, as he understands it, presupposes dualism of this kind since its essence lies in enframing.

This explanation of how Heidegger arrives at his view of technology as an offshoot of metaphysics ought not to conceal the basically abstract, and therefore fundamentally incomplete, nature of his view. Whatever else it is, technology is not called into being through the problem of metaphysics. Heidegger's occasional suggestion that it is Being which lurks behind and affects technology is evidently mythological. For it is clear that technology, including modern technology, was called into being by human beings confronted with specific tasks arising within a specific social and historical milieu. There is no indication in "The Question Concerning Technology" or in Heidegger's other writings that he has a real comprehension of the nature of society or of its historical evolution. In *Being and Time*, where he was clearly under the influence of Kierkegaard, his understanding of the social as being-with was mainly negative, depicted as an inauthentic mode of life which one must leave behind. In later writings, there is no evidence that he has made progress in comprehending the social context. In this respect, he is clearly surpassed by Sartre, his most important French student. In his later writings, Sartre went beyond the abstract analysis he presented in *Being and Nothingness* to offer a concrete analysis of society in terms of such concepts as practice and lack.

Heidegger did not have a theory of society and, hence, cannot integrate his view of technology into a wider understanding of the social context. As a result, his theory of technology fails on at least two counts. First, he simply is unable to grasp the origins of technology from a historical perspective, which in turn leads him to attribute these origins to an extension of metaphysics. Second, he is unable to comprehend the social role of technology, which he also attributes to Being. It is clear that technology arose to satisfy human needs and desires. But Heidegger, who only scrutinizes technology within the context of his deeper concern with metaphysics, seems incapable of grasping the relation of technology to society and human being.

Because he cannot grasp the relationship of modern technology to the modern social world, Heidegger fails to comprehend the manner in which technology is not beyond human control but at least potentially subject to the wishes of human being. As soon as we admit that our actions often have consequences beyond our intentions, we can comprehend that there is something about history which escapes human control. But even if we acknowledge that in principle there will always be something about the historical process which will not merely coincide with the men and women whose actions constitute it, it does not follow that technology as such is beyond human control. That is a leap which no rational argument seems to require, but which Heidegger is apparently willing to make, because he has presupposed the action of a transpersonal reality like Being which, he believes, communicates with us through technology.

The deeper question is the extent to which technology is or can come under human control. I have already indicated why I hold that we must reject the claim that the technology we consciously employ for our own ends is something that lies under the control of a transpersonal reality. It is difficult to deny that technology sometimes seems to serve human beings. Obviously, there is an important distinction between one or another specific form of technology and technology in a general sense. Perhaps the best way to understand the idea of technology in general is as one aspect of modern social life. The history of mankind records the continued striving of people and groups to free themselves from natural and self-created constraints in order fully to develop their individual capacities. Whether human beings will finally be successful in this effort, whether we will ever free ourselves from even self-imposed constraints, is part of the larger experiment of human history whose outcome no one can know in advance. We cannot rationally assert, Heidegger notwithstanding, that man will never be able to bring technology under control. We may at least hope that such control will occur.

My conclusion is that Heidegger's reading of technology is deeply and irremediably flawed since he fails to identify basic elements of technology and conflates the phenomenon in general with technique and art. It is no accident that his reading of technology is flawed. The flaw follows directly from his profoundly antimodernist perspective. For Heidegger, the way to understand the world in which we live is through the elaboration of an alternative story of the meaning of Being. Such a story rejects the story mainly embraced since the pre-Socratics as one that arises from a false turning early in the path, a divergence from another, truer turning which would continue in the correct way. As in his analysis of modern thought, so in his analysis of modernity Heidegger proceeds from the assumption that what we think and what we are can be redeemed only by returning to an earlier and truer view of things. Hence, it is hardly surprising that he cannot grasp the way in which technology arose and functions within modernity since modernity itself, or at least what we mean by this term, functions for him as a fall away from an earlier and truer way of being, and perhaps even as a fall away from Being. I submit that neither Heidegger nor anyone else can comprehend modernity or its elements—including technology—unless these elements are approached in their own right without preconditions and allowed to tell their own story. Certainly, we are entitled to expect nothing less from any form of philosophy, including phenomenology.

The anti-democratic nature of Heidegger's thought is visible in his anti-anthropological conception of technology. As a political theory, democracy in all its forms necessarily presupposes a general conception of human being. In the course of the turning in his thought, or *Kehre* as he terms it, Heidegger turned away from the anthropological perspective which underlay his earlier approach to Being. His writings after the turning presuppose an anti-humanist perspective clearly described in his "Letter on Humanism." But a theory which goes beyond an anthropological concept is incompatible with any meaningful view of democracy which surely must depend on such a concept.[30] His anti-metaphysical theory of technology from the post-anthropological viewpoint of his new thinking is by definition non-democratic and even anti-democratic since it presupposes as its defining condition the rejection of the anthropological point of view which is the foundation of democracy. It is, then, no accident that Heidegger's fundamental ontology led him to embrace Nazi totalitarianism since his conception of Being is incompatible, and was understood by him to be incompatible, with any form of democratic theory.

Duquesne University

NOTES

1. See Victor Farias, *Heidegger et le nazisme* (Paris: Editions Verdier, 1987).
2. Lukács writes about Heidegger in different ways in a series of books. For the analysis of Heidegger's relation to Nazism, see especially *Existentialismus oder Marxismus?* (Berlin: Aufbau-Verlag, 1951), *passim* but particularly "Anhang: Heidegger Redivivus," pp. 161-183.
3. See, for instance, the recent, equivocal study by Jacques Derrida, *De l'Esprit* (Paris: Editions Galilée, 1987).
4. For this excessive judgment, see the discussion of Heidegger and Jaspers in *The Destruction of Reason*, trans. Peter Palmer (Atlantic Highlands, N.J.: Humanities, 1981), chapter 4, part 6: "The Ash Wednesday of Parasitical Subjectivism," pp. 489-521.

5. See Martin Heidegger, *An Introduction to Metaphysics*, trans. Ralph Mannheim (New Haven: Yale University Press, 1977), p. 199. For the German-language original version of the statement, see Martin Heidegger, *Einführung in die Metaphysik*, in *Gesamtausgabe* (Frankfurt: Klosterman, 1983), volume 40, p. 208: "Was heute vollends als Philosophie des Nationalsozialismus herumgeboten wird, abermit der inneren Wahrheit und Grösse dieser Bewegung (nämlich mit der Begegnung de planetarisch bestimmten Technik und des neuzeitlichen Menschen) nicht das Geringste zu tun hat, das macht seine Fischzüge in diesen trüben Gewässern der 'Werte' und der 'Ganzheiten'." For Heidegger's misleading effort to suggest that this statement had merely tactical value in order to defend freedom of speech, see p. 233, where in a letter to S. Zemach, in a comment on this passage, he writes: "Die verständigen Hörer dieser Vorlesung haben daher auch begriffen, wie der Satz zu verstehen sei. Nur die Spitzel der Partei, die—wie ich wusste—in meiner Vorlesung sassen, verstanden den Satz anders, sollten es auch. Man musste diesen Leuten hie und da einen Brocken zuwerfen, um sich die Freiheit der Lehre und Rede zu bewahren."

6. See *Existentialismus oder Marxismus?*, p. 19.

7. *Ibid.*, p. 28.

8. See John D. Caputo, "Demythologizing Heidegger: *Aletheia* and the History of Being," *Review of Metaphysics* 41 (March 1988):542.

9. The best treatment of this problem of which I am aware is Jean Grodin, *Le Tournant dans la pensée de Martin Heidegger* (Paris: Presses Universitaires de France, 1987).

10. "Letter on Humanism," in Martin Heidegger, *Basic Writings*, ed. David Farrell Krell (New York: Harper and Row, 1977), p. 215.

11. See Heidegger, *Introduction to Metaphysics*, p. 199.

12. See Pierre Aubenque, "Encore Heidegger et le nazisme," *Le Débat*, 48 (January-February 1988):120.

13. See "The Question Concerning Technology," in Heidegger, *The Question Concerning Technology and Other Essays*, trans. William Lovitt (New York: Harper and Row, 1977).

14. Martin Heidegger, *Being and Time*, trans. Macquarrie and Robinson (New York: Harper and Row, 1962), p. 409; *Sein und Zeit* (Tübingen: Niemeyer Verlag, 1963), p. 358.

15. Heidegger, *Being and Time*, p. 100; *Sein und Zeit*, p. 70.

16. Heidegger, *Introduction to Metaphysics*, p. 38.

17. Heidegger, *Question Concerning Technology*, p. 116.

18. Heidegger, *Basic Writings*, p. 220.

19. Heidegger, "Letter on Humanism," in *Basic Writings*, p. 239.

20. Heidegger, *Question*, p. 31.

21. The clearest statement of Heidegger's view of enframing of which I am aware is his characterization of *Machenschaft* as *Der Bezug der Unbezüglichkeit*. This expression can be rendered very approximately as the unframed frame—in short, as a notion of horizon. See Heidegger, *Beiträge zur Philosophie (das Ereignis)*, ed. F. von Hermann (Frankfurt: Klosterman, 1989).

22. Heidegger, *Question*, p. 19.

23. Heidegger, *Being and Time*, p. 438; *Sein und Zeit*, p. 386.

24. Heidegger, *Question*, p. 26.

25. *Ibid.*, p. 24.

26. See Karl Marx, "The Eighteenth Brumaire of Louis Bonaparte," in Marx and Friedrich Engels, *Basic Writings on Politics and Philosophy*, ed. L. Feuer (Garden City, N.Y.: Doubleday, 1959), p. 320.

27. Heidegger, *Question*, p. 24.

28. See the denial that man controls technology in Heidegger, *Identity and Difference*,

trans. J. Stambaugh (New York: Harper & Row, 1969), p. 34.

29. See Frederick Ferré, *Philosophy of Technology* (Englewood Cliffs, N.J.: Prentice-Hall, 1988), p. 26.

30. Heidegger emphasizes the anti-democratic nature of his thought in the *Spiegel* interview in the famous claim that only a God can save us. See "Only a God Can Save Us: Der Spiegel's Interview with Martin Heidegger," *Philosophy Today* (Winter 1976). For an earlier version of this claim as the idea that only *das Deutsche* can save us, see Heidegger, *Heraklit*, ed. M. Frings (Frankfurt: Klosterman, 1987), pp. 107-108 and 123.

ALBERT BORGMANN

THE MORAL ASSESSMENT OF TECHNOLOGY

Modern technology is the unacknowledged ruler of the advanced industrial democracies. Its rule is not absolute. It rests on the complicity of its subjects, the citizens of the democracies. Emancipation from this complicity requires first of all an explicit and shared consideration of the rule of technology.

Assessments of this sort are undertaken today in various ways, none of them having decisive force in the shaping of democratic policies. Hence much energy and good will is spent on getting a better hearing for the consideration of technology. While no reasonable person would counsel a moratorium on these commendable endeavors, it is still important to pause and reflect on the variety of assessments that technology is subjected to, to ask how they hang together and how radical their force finally is. This is an important task even if one does not accept the way it is usually done. One way of advancing this task is a second order approach, one of assessing the ways in which technology is commonly assessed. Such an approach has the virtue of circumspection. To avoid the concomitant vice of blandness, a substantive concern must finally guide the investigation.

I want to introduce this concern in two steps. The first leads us to a moral assessment of technology. As Dan W. Brock reminds us, there is something like "the finality, overridingness or authority of morality: moral judgments (reasons, rules) override non-moral judgments in cases of conflict."[1] Focusing on the moral assessment of technology provides, therefore, a first specification of the task at hand.

Further specification is needed, however, since there is no common agreement on the standards that are to guide a moral assessment of technology. This second step can be taken in a substantive way only, i.e., by saying concretely what endangers and advances human welfare. Such a resolution is a first order enterprise; it deals directly with the object of the good life.

More specifically, I will consider four assessments of technology, the technical, the modern, the traditional, and the quotidian. This sequence can be generally characterized in three ways. First, it constitutes an order of increasing sufficiency. Each of the four kinds of assessment will be seen to be insufficient from the standpoint of its successor.

Second, the sequence is one from a nonmoral or weakly moral assessment (the technical) to moral assessments (the last three). Third, it is a sequence of second order to first order inquiries. I will consider the technical, modern, and traditional assessments predominantly from a second order point of view and then engage the fourth, the quotidian assessment, more or less immediately.

But before I turn to the discussion of these various assessments, I must say a brief word about their object, the thing to be assessed, namely, technology. What is technology? In the debates of social and philosophical critics, three senses of technology usually come into play: technology as an ensemble of powerful but value-neutral instruments; technology as an overwhelming and determining force; and technology as the form of life that is paradigmatic of the modern era. These may be called the instrumental, the deterministic, and the paradigmatic senses of technology. It is the third sense, technology as our paradigmatic form of life, that I will be assuming here.[2]

Turning, then, to the technical assessment of technology, I note that it is the most familiar, least conspicuous, and most characteristic of technology itself. Theoretically seen, it employs the syncategorematic or rational sense of good. In asking, for example, whether a Rolls Royce is a good car, it does not apply an independent criterion of goodness, but allows the notion of the modern automobile to infuse "good" with content and significance. Speaking more formally, we can say that the technical assessment of a Rolls Royce asks: Does a Rolls Royce have the properties to a higher degree than an average or standard car which it is rational to want in a car, given what cars are used for?[3] As it stands, the syncategorematic or rational application of goodness to assess technology suffers from excessive generality. It has no particular bearing on technology. A common way of restricting rational goodness to technology is to specify it as efficiency. This move figures importantly in Jacques Ellul's *Technological Society*. In his view, efficiency is the best adaptation of available means to a desired end.[4] This criterion has in common with rational goodness its optimizing character and its syncategorematic incompleteness. Beyond that it has the merit of highlighting the unique importance of the means-end relation in technology.

Still, it too suffers from generality and incompleteness. Hence others have sought to spell out efficiency more definitely. Adapting the comments of Henryk Skolimowski, we can say that various branches of technology are guided by characteristic notions of efficiency, surveying by accuracy, civil engineering by durability, mechanical engineering by the yield of an engine, architecture by aesthetics and utility.[5] On a much grander scale, David Billington has developed a notion of structural art as the guiding ideal of civil engineering. He explains structural art as the pursuit of efficiency in the use of materials, of economy of cost, and of elegance of form.[6]

These various criteria are characteristic of technology as a form of life in two ways. For one, they seem eminently reasonable and appropriate. We recognize in them the guideposts that are actually being followed in the construction of the technological universe. For another, however, they conceal even more than they disclose. The application of the criteria of efficiency secures technological structures that are individually or locally benign but diverts attention from the larger questions of justice and safety.

Dissatisfaction with this concealment has led to a moral assessment of technology. It has been urged that what seems beneficial here and now is not always so in a larger perspective. The application of a pesticide may insure good harvests this year and even the next several years. But it also leads to an injurious accumulation of toxins in animals and humans. The use of a spray may be safe and convenient in patching up a scratch on a car; but if chlorofluorocarbon compounds drive the spray, they also contribute to the destruction of the protective ozone shield in the upper atmosphere. And finally a rocket that seems a marvel of effective defense, considered by itself, is in fact a component of the rising tide of destructive nuclear terror.

What I am illustrating here is the powerful and widespread critique of the failure of technological safety advanced by environmentalists and peace activists. Hans Jonas has given this movement its most eloquent philosophical expression in *The Imperative of Responsibility: In Search of an Ethics for the Technological Age*.[7] This concern for safety has an analogue in the concern for justice. What is beneficial for the upper and middle classes of the advanced industrial countries is lacking or detrimental for the poor of the First, Second, and Third Worlds. Liberals such as John Rawls, Ronald Dworkin, and Lester Thurow have powerfully scored the injustice of the prevailing common order in the United States and, by implication, of the industrial democracies.[8]

If technology is our paradigmatic form of life, then environmental, pacific, and liberal critics of the prevailing common order must be critics of technology as well. But most of these critics would reject this inference since they hold to an instrumental sense of technology. This is not just a terminological reservation. Most social and moral critics do not believe that the ensemble of technological artifacts immediately expresses and all but enforces a distinctive and morally charged form of life. They believe that moral codes and principles are articulated autonomously and that typically their authority is exercised mediately through the proper use of technological devices not immediately in the very conception and construction of a technological artifact.

Thus the environmental and liberal assessment of technology, while critical of the use of technology, is still consonant with the notion of technology as a neutral, though powerful, instrument.[9] It criticizes a particular employment of technology but leaves the intrinsic moral quality of technological devices unexamined and implicitly sanctions it. It does so because it has the same ancestry as technology. Both technology and its critics are the progeny of the Enlightenment, of its promise of liberty and prosperity. The environmental and liberal critique of technology amounts to a call for the recovery and fulfillment of the promise of technology, not to a searching examination of that promise.

One can therefore call this sort of critique the modern moral assessment of technology. What is formally called technology assessment in the United States and undertaken by the congressional Office of Technology Assessment in fact undertakes such morally modern and technical assessments of various technologies. But just as the technical assessment of technology is found to be wanting from the modern moral point of view, so the latter has recently been criticized from the standpoint of traditional morality.

Like the term "modern," the notion of "traditional" morality is to be taken here in a special sense, exemplified by the work of Alasdair MacIntyre and of Robert Bellah and his associates.[10] The critique of these scholars goes further than that of the moderns because they put into question the morality of the modern temperament. They find it to be shallow, unsettled, and impoverished. What is relevant for the moment is merely the radicality of their charge. These critics are not trying to recover the integrity or comprehensiveness of the standard modern project; they are casting doubt on its intrinsic moral quality.

I am essentially in agreement with their critique; and so the question whether the critique is well-founded is one that I ought to answer myself. The essential part of the traditional assessment that I share regards the losses that it describes and the redemptions that it urges, both the losses and redemptions centering on the classical virtues in the case of MacIntyre and on the biblical and republican traditions in the case of Bellah.

But MacIntyre's project needs the correction and Bellah's the complement of a close study of the material culture that we live in. These traditional moral critics consider predominantly the ways in which we reflect and talk about our lives and the ways in which we act insofar as our actions are independent of our material surroundings.

Such considerations are certainly needed for a moral assessment of the contemporary human condition. But are they sufficient? Do they

THE MORAL ASSESSMENT OF TECHNOLOGY 211

consider the full scope of our moral condition? If technology is the characteristically modern form of life, then an incisive critique of modernity must be one of technology as well. In MacIntyre, technology does come into view under the headings of the therapeutic attitude, bureaucratic individualism, and goods that are external to practices.[11] But MacIntyre pays no attention to the concrete and quotidian circumstances that specify and sustain attitudes, bureaucracies, and practices. In Bellah's book technology in all its dimensions is very close to the surface. But even there, and much more so in MacIntyre, closer consideration of the material culture would have disclosed more vividly major debilities in our prevalent form of life.

What is finally needed, then, is a moral assessment that takes into account the concrete dailiness that channels our endeavors and aspirations. This may be called the quotidian moral assessment of technology, and to establish its significance the best I can do within the limits of this essay is to provide a suggestive illustration of it.

There is a pervasive uneasiness in the American culture about the status and influence of television. Social scientists have done much to explore and explicate this sentiment. But in MacIntyre's book television is of no concern. In Bellah's book it does receive attention, but mostly in regard to the values it presents and advocates, not as regards the values or the form of life it constitutes and represents.[12]

Television is a way of consuming reality. To consume in the prevailing contemporary sense is to appropriate or ingest something in the imperious and unencumbered way that depends on the sophisticated and powerful machinery of modern technology. The latter makes the riches of the world available to us as commodities that are entirely subject to our desires. In technology, machinery and commodity are divided from and related to one another as means and ends in a radically novel version of this relation. Never before have ends been so unencumbered by means and hence so conveniently available. Nor have means ever been so inaccessible and unintelligible to the ordinary citizen.

Television is perhaps the paradigmatic instance of this new relation and of the approach to reality informed by that relation. Through television the events of the world, the treasures of culture, the secrets of nature, the dramas of history, the pleasures of sex are all laid out before us and available at the pushing of buttons or the turning of dials. This is the paradigm of unencumbered and abundant ends, procured by means of a powerful, sophisticated, and concealed machinery.

To have such unrestricted command of the world is, in one sense, the fulfillment of the dream of liberty and prosperity. That we are uncertain of the fulfillment is apparent from our unceasing determination to ignore the fulfillment and to revive the dream, to dream up ever more advanced technologies able to procure the world for our consumption.[13] Underneath this restless dreaming there is a dawning realization that consumption as a central feature of our life is more debilitating than liberating and enriching.

It is debilitating because a world reduced to consumption goods or commodities no longer engages us in the depth of our bodily, mental, and communal capacities. There is a double and symmetrical loss. The world, when reduced to consumable commodities, has been gutted and trivialized. Humans, when reduced to consumers, have become solitary and passive.[14]

If this suggestion has merit, then there are two limits to the traditional moral assessment of technology. The first is the lack of appropriately critical categories and vocables. What is wrong from the traditional point of view with television as a prominent feature of our way of life? Traditional morality will take issue with this and that in television. But as a whole, as a way of appropriating the world, television has for the most part escaped the judgment of traditional morality. What is needed is a new vocabulary of criticism that contains notions such as disengagement, reduction, trivialization, consumption, commodity, procurement, and more.

Such vocables are needed to articulate our misgivings about television in a measured and constructive way. Traditional and current terms of moral assessment are surely unhelpful in this regard. Television watching is neither a sin nor a crime. It is neither socially unjust nor environmentally destructive. But it disengages us from one another; it reduces us to passivity; it trivializes the sacred; it replaces celebration with consumption and challenge with commodity; and it procures for us what we should strive for and come to deserve.

The second limit of traditional morality follows from the first and lies in the fact that without a penetrating critique of the quotidian status quo the traditional proposals will seem futile or irrelevant. MacIntyre counsels a return to practices and virtues. What stands in the way of such proposals is less the supposedly chaotic state of moral discourse he bemoans than a deeply entrenched life of production and consumption that is not so much opposed as it is insensitive to and incapable of virtues and engaging practices. Traditional morality must on its part engage and open up the quotidian condition of technology and spell out its proposals in concrete and fruitful contrast to the prevailing circumstances.[15]

I said earlier that the four assessments of technology that I have been considering are arranged in an order of increasing sufficiency. As is clear now, this does not mean that each assessment is more sufficient than its predecessor; each rather adds a necessary consideration to those that precede it. Moral assessments cannot dispense with technical ones; and the concern with virtue should not replace the concern with justice. But in every case, the latter assessment needs to complement and guide the former.

University of Montana

NOTES

1. Dan W. Brock, "The Justification of Morality," *American Philosophical Quarterly* 14 (1977): 71.
2. For more detailed discussion see my *Technology and the Character of Contemporary Life* (Chicago: University of Chicago Press, 1984), pp. 7-12 in particular.
3. This is adapted from John Rawls, *A Theory of Justice* (Cambridge, Mass.: Harvard University Press, 1971), p. 399.
4. Jacques Ellul, *The Technological Society*, trans. John Wilkinson (New York: Random, 1964), 21 and passim.
5. Henryk Skolimowski, "The Structure of Thinking in Technology," in C. Mitcham and R. Mackay, eds., *Philosophy and Technology* (New York: Free Press, 1972), pp. 46-49.
6. David P. Billington, *The Tower and the Bridge* (New York: Basic Books, 1983).
7. Hans Jonas, *The Imperative of Responsibility* (Chicago: University of Chicago Press, 1984).
8. Rawls, *Theory of Justice*; Ronald Dworkin, "Why Liberals Should Believe in Equality," *New York Review of Books* 30 (3 February 1983): 32-34; Lester C. Thurow, "A Surge in Inequality," *Scientific American* 256 (May 1987): 30-37.
9. This is only roughly true. The "deep ecology" movement, e.g., has fundamental reservations about technology.
10. Alasdair MacIntyre, *After Virtue*, 2d ed. (Notre Dame, Ind.: University of Notre Dame Press, 1984). Robert N. Bellah, Richard Madsen, William M. Sullivan, Ann Swidler, Steven M. Tipton, *Habits of the Heart* (Berkeley: University of California Press, 1985).
11. MacIntyre, *After Virtue*, pp. 30-31, 35, 71, 187-191.
12. Bellah, *Habits*, pp. 279-81.
13. See, e.g., Hans Fantel, "Enhanced TV Standards Still Elusive," *New York Times* (11 January 1987) section 2, p. 30; and the contributions to "The Seer's Catalog" by Bill Gates and Tony Verna in *Omni* 9:4 (January 1987): 38 and 40.
14. We remain active and engaged with the world through labor. But this tie to reality is suffering through technology as well. See my *Technology and Contemporary Life*, pp. 114-124.
15. For such proposals see my "Communities of Celebration: Technology and Public Life," in F. Ferré, ed., *Research in Philosophy and Technology*, vol.10 (Greenwich, Conn.: JAI Press, 1990), pp. 315-345.

JOSEPH MARGOLIS

POLITICAL MORALITY UNDER RADICAL CONDITIONS

I

When, in the seventeenth century, the English philosopher Thomas Hobbes defended absolute monarchy, his masterpiece *Leviathan* was perceptively received with a great deal of suspicion and dissatisfaction. For it was at once clear—Hobbes drew attention to the fact—that he had not defended the divine right of kings and his concessions to parliamentary interests were, however conceptually serious, utterly lacking in any explicit provision for parliament's political power. The boldest stroke of Hobbes's endeavor lay in collapsing the distinction between egoistic prudence and political morality. Nevertheless, in achieving that extraordinary simplification, Hobbes never doubted that "all men by nature reason alike, and well, when they have good principles" and that the "signs of prudence are all uncertain" though "the signs of science are some, certain and infallible; some, uncertain."[1] For Hobbes, human nature was essentially invariant, and the unreliability of prudence lay with its depending on contingent and particular experience, which obscured (rather than precluded) our grasp of the laws of nature (the laws of prudence), as well as with its contrast to science proper (the theoretical knowledge, in effect, of causal invariances).

Hobbes took the enormously important step of insisting on the artifactual nature of politics, political morality, law, and morality in general; but he did so, curiously, by insisting on the natural and unchanging nature of man and human reason. There is that much of a gulf between Hobbes's seventeenth-century confidence and the most characteristic twentieth-century theorizing. For, in our own time, all speculation that may be called postmodernist, poststructuralist, antimodernist—also, all speculation dissatisfied with any simple disjunction between modernist and postmodernist alternatives—is committed to treating human reason and human nature as artifacts *of* the societal, historical, praxical, or technological conditions under which the members of *Homo sapiens* come to exhibit the functions of full-fledged persons. Actually, in our own time, even perception and the laws of nature are treated without difficulty as artifacts;[2] and so, the very idea of a stable *natural* science or practice of law or political morality claiming normative validity or legitimation is often viewed askance. Still, it is difficult to construe the achievements of any of these disciplines entirely in terms of mere indoctrination.

215

The point is that Hobbes's daring collapse of morality and prudence obscured the fact that he also confirmed the fixities of human nature in terms of which the disjunction between theoretical and practical reason could still be maintained. Prudence remains, for Hobbes, causal knowledge ratiocinatively applied to man's passions and appetites. There is, for Hobbes, no theoretical knowledge of good and evil in Aristotle's sense, for good and evil are not independent features of an independent world: a theoretical science separated from prudential concern remains, however, entirely possible for Hobbes. Even the skeptical attenuation of science in terms of Hobbes's nominalism does not erase the epistemic distinction of a theoretical science. Of course, Hobbes's nominalism happens to be a primitive and inadequate conjecture, since nominalism is quite incapable, in principle, of epistemic application to new cases going beyond the exemplary specimens to which the relevant predicates are first applied. The only possible account of general terms congruent with theories that regard the intelligible world as, in some essential sense, both artifactual and cognitively intransparent is an account of a conceptualist sort strengthened in the direction of the realism of our artifactualized world; for then, the conceptual powers of the mind are adequated to what we take the real properties of the cognized world to be. Quite clearly, such an account is particularly perspicuous for realist views of human culture and for all internal realisms applied to physical nature.[3] Equally clearly, all such theories are bound to feature the epistemically formative or preformative function of technology and *praxis*. A conceptualist account of thinking is notably hospitable to the right mix of the realist and the artifactual under the constraint of a constructive theory of persons.

For us, in the late twentieth century, the emphasis on the preformative forces of historical existence affects the cognitive prospects of science every bit as much as politics and law and morality. Hence, the denial of a disjunction between theoretical and practical reason does not, for us, require (though it need not disallow) the failure of prudence and morality—unless it also requires the failure of science. In our own time, the mark of modernity rests with *the holistic influence of an entire enculturing society acting on the aptitudes of its aggregated members so as to shape a cognizant order, at once scientific and prudential, for their own lives.* For, in our sense, currently, we and our cognized world are artifacts of our own society's enculturing processes: we reconstitute ourselves, our society, our world, our history again and again by an active intervention that is ineluctably praxical even at its most theorizing.[4] (This is not, of course, to treat the world as a fiction or as infinitely plastic, only to reject a clear disjunction between realist and idealist ingredients in any would-be science.)

Michel Foucault may have supplied the most memorable (if extreme) short instance of this distinctly contemporary point of view when, in *Les Mots et les choses*, anticipating by nearly a decade his own genealogical studies in *Surveiller et punir*, he asserts:

> Before the end of the eighteenth century, *man* did not exist—any more than the potency of life, the fecundity of labor, or the historical density of language. He is a quite recent creature, which the demiurge of knowledge fabricated with its own hands less than two hundred years ago; . . . there was no epistemological consciousness of man as such. The Classical *episteme* is articulated along lines that do not isolate, in any way, a specific domain proper to man.[5]

Karl Marx, of course, had already construed all the imputed structures of the intelligible world, including those structures imputed to man himself, as historicized constructions generated within the actual forms of life of particular societies. That is what, transforming Aristotle's notion of man's social and political nature,[6] Marx identified as the praxical within human theory and human existence. Foucault goes well beyond Marx in a Nietzschean sense, in that the normalizing *epistemes* of particular phases of *praxis* are themselves construed as the unaccountable congealings of inchoate forces (more subterranean than *praxis* itself) that cannot be assigned to any merely prior *epistemes*.

In short, Foucault introduces a radical discontinuity in the succession of *epistemes* that is not explicable, as it would have been for Marx, in terms of the emergent possibilities of the specific structured and structuring processes of particular societies. Nevertheless, what Foucault succeeds in confirming by this maneuver is at least the radical contingency, the intransparency, the artifactualization, of human life and understanding. In that sense, Foucault's vision *is* the postmodernist analogue (perhaps the clearest) of the primacy of *praxis* and technology favored in the modernist (preeminently, the Marxist) conception of science and practical life. Schematically, then, Marx and Foucault remain Hobbes's principal—distinctly radical—successors: first, in historicizing human nature as a function of technology and practice (Marx's modernist conception of a historicized science of man); second, in genealogizing all historicized discoveries (Foucault's postmodernist transformation of the science of human history).

Within the space of either vision, of course, the possibility cannot be denied that, at any given level of discourse—at the level of logic and mathematics, or of physics, or biology, or sociology, or political morality—conjectured invariances may still be saliently isolated that effectively escape the naive epistemic optimism of the modernist as well as the anarchical pretentions or threats of the postmodernist. Science and prudence seem to change so conservatively that they still yield plausible notions of rationality, legitimation, purposive inquiry and commitment; they still sustain disputes (both first-order and second-order) about the compared promise of alternative cognitive and active claims. Only now, their pertinent invariances can no longer be drawn from theoretical sources that escape the contingencies conceded. We are confined, at least initially, to first-order invariances—to "indicative" invariances, as we may call them—salient, provisional, horizonal, perspectival, plural, interest-bound, ranging over both normative and non-normative regularities. (For example, in our own time, the Western world is distinctly but not exclusively inclined to view human nature, both descriptively and normatively, in terms of economic man.) Our adjustment propels us beyond Hobbes but in a way already implicated in Hobbes's own bold conjecture. Furthermore, in doing that, it settles the sense in which *praxis* and technology remain the nonprivileged sources of our account of ourselves and our world. That is the point of *our* reading of *Hobbes's* having conflated prudence and morality: the essential point of any reading of man in a world now permanently altered by the French Revolution.

II

Once we collapse the distinction between theory and practice—essentially by some combination of second-order claims affirming: (i) the cognitive intransparency of the real world; (ii) the artifactualization of the intelligible world *and* of our own self-understanding; and (iii) the historicized preformation of human inquiry and activity—the perceived limitations of science and prudence may reasonably be expected to affect each of those domains, from the direction of the other. For example, the artifactual nature of nomological invariances within the physical sciences is bound to affect adversely the very idea of objectively legitimated universal moral principles. Of course, Hobbes did not consider that his own famous laws of nature (his prudential laws) would prove to be mere formulations of an emerging bourgeois ideology: once we perceive that possibility, however, we can hardly resist the further charge that the would-be invariances of the physical sciences also betray an ideology-like dimension. Roberto Unger has caught this possibility very well when he cannily observes that "It is always hard for the naturalist [the natural scientist positing would-be nomological invariances] to tell which aspects of the developed natural world are due to the particular divergent sequences that have occurred [in the shifting history of a science] and which aspects hold, no matter what the sequence."[7]

Nevertheless, in the physical sciences, the strong objectives of prediction, explanation, successful technological intervention based on causal regularities do provide compelling grounds for minimizing the charge of mere ideology and for maximizing the claims of realism even under conditions (i)-(iii). In fact, it is precisely because the objectives of the physical sciences *are* pursued within those constraints, and because the theories that best satisfy them there *are*, increasingly, radically underdetermined by the perceptual data by which they are tested, that *we* are bound to construe our own second-order efforts at legitimation as rationally unavoidable and as never merely prejudiced or horizonal. We may not be able to assign objective measures of objectivity to specifically separate competing programs of inquiry; but at least holistically—pragmatically, blindly—human inquiry cannot be said to fail to grasp a significant part of the real structure of the natural world. The underdetermination of theory *and* the dependence of the success of predictive observations on felicitously selected theoretical unobservables viewed in the light of the remarkable success of the physical sciences, demonstrates (as well as any argument could): first, that we cannot afford to ignore second-order legitimative questions regarding the reasonableness of pursuing this inquiry or that; and second, that positive answers to such second-order questions cannot convincingly be said to violate such constraints as those of (i)-(iii), already collected, or to ignore the consequences of being bound by such constraints.

Considerations of these sorts expose the extreme carelessness of so-called postmodernist arguments against the continuation of legitimative or transcendental philosophical concerns, wherever the latter honor the sort of constraint already tallied.[8] The point is that the loss of cognitive transparency and privilege does *not* in the least entail the loss of objective second-order claims—or the irrelevance of such claims. Just the contrary: the objectivity assignable within first-order science entails a range of objective second-order inquiries. Empirical underdetermination in sciences—particularly when it is seen to be linked to possible consequences of the greatest human importance (global starvation, for example)—obliges us *rationally* never to abandon legitimative questions. *There is no plausible argument for denying their relevance.* The appropriate question concerns what we may make of our cognitive resources.

Given the denial of a principled disjunction between theory and practice, our argument, pointedly drawn from the natural sciences, provides a basis for entertaining the further possibility that political, legal, and moral arguments may similarly be legitimated—though, if so, *not* in the manner of the sciences (since, contrary to Hobbes, causal questions can form only the least distinctive part of normative inquiries) and probably *not* in any way disjoined from prudential considerations (contrary, for instance, to Kant's notorious view).

In fact, if Hobbes had only subscribed to constraints (i)-(iii), then his own denial of a disjunction between political morality and prudence would have been tantamount to a denial of a disjunction between theory and practice; and then, the recovery of an objective science *within* ideological contingencies would have demonstrated (so far at least) that an objective politics, morality, or law would not be entirely out of the question. Put more promisingly, the idea of objective norms and judgments regarding human action and human commitment could be grounded in part in the objectivity of causal inquiry; could escape mere prejudice, mere rationalization, mere circularity, mere self-serving interest if the prudential concerns of the race could in some measure be objectively determined; and could be provided with a plausible rule of objectivity distinctly suited to the special features of discourse about human life and behavior. This last provision does not signify a new disjunction between theory and practice, only a recognition of the fact that the human world exhibits features (intention, purpose, history and the like) that do not appear within physical nature narrowly construed.

There is, therefore, every reason to suppose that, *if* physicalism fails,[9] *if* the difference between persons and physical objects is conceded, *if* cognitive transparency is denied, the physical sciences and practical reason would still have to be assigned an entirely different character from one another. (They need not, however, and cannot, possess different cognitional sources.) In principle, the objectivity of politics, law, and morality *could not* be as resistant to the current of historical bias as the physical sciences. For one thing, they are ineluctably tied to the affirmable interests or desires of men (indifferently characterized as subjective or prudential at this point); and for another, the normative questions they pose cannot, in principle, be reduced to causal questions (as Hobbes fully appreciated). Seen in this way, Hobbes's solution may be recovered as an optional specimen of what an objective political morality might look like *relativized to seventeenth-century concerns*. It would remain a naive solution in supposing human nature to be transhistorically invariant, in supposing that some unique, ideally adequate account of human nature *could* be formulated, and in supposing that Hobbes had actually managed to capture the essential features of objective prudence.

Nevertheless, we sense here the prospect of formulating a distinct notion of the "objectivity" of first- and second-order moral, legal, and political claims, in spite of the intransparency and artifactualization of ourselves and our world and the preformational prejudices that affect our inquiries and commitments. It may then be no more than a prejudice, rather like the prejudice of the unity of science (that insists that the study of man must conform to the canons for the study of inanimate nature), that persuades us that any supposed objectivity ascribed to political, legal, and moral discourse must derive from the canons of objectivity suited to the sciences.[10] There is no possibility of satisfying such a condition, and it hardly makes sense to insist on it in a world accessible only through a historicized *praxis* and technology. Either we must give up the pretense of an objective morality or politics, or else we must construe it in a radically different way.

Consider, now, the postmodernists, for they reject altogether or pose radical difficulties for any form of second-order legitimation.[11] Richard Rorty, for example, offers a rationale for loyalty to "us postmodernist bourgeois liberals . . . to suggest how such liberals might convince our society that loyalty to itself is morality enough, and that such loyalty no longer needs an ahistorical backup . . . . I would argue [says Rorty] that the moral force of such loyalties and convictions consists *wholly* in this fact [the fact of such loyalty], and that nothing else has *any* moral force."[12]

At least three distinct difficulties suggest themselves that Rorty nowhere addresses: first, he appears to argue (he certainly argues elsewhere) that if "loyalty" to a given morality (or to any other cognitive or normative code or cause) could be legitimated, it would have to be legitimated "ahistorically," that is, by way of some presumed cognitive privilege; second, he does not show why legitimation within historicized, intransparentist constraints is impossible or irrelevant; and third, he does not show why "loyalty" of the sort *he* supports could not be reasonably judged to be defective or disordered (say, by analogy with colorblindness or by the example of the Nazis), or why his own advocacy is not itself an attempt at legitimation within historicized constraints—very probably a disguised and peculiarly uncompelling attempt of that kind. The bare mention of these difficulties constitutes a *reductio ad absurdum*.

A second postmodernist specimen may be collected from Michel Foucault's analysis. Particularly in his later years, Foucault emphasized the priority of the ethical or practical over the cognitive or theoretical. Foucault stessed "the way a human being turns him- or herself into a subject" by internalizing the "objectivizing" modes of a society, by internalizing its "dividing practices" by which (as he says) "the subject is either divided inside himself or divided from others" (as, perhaps, mad and sane, sick and healthy, criminals and "good boys").[13]

"The main objective of these struggles [Foucault asserts, that is, struggles and forms of resistance within and against a controlling society—against one's own, in fact] is to attack not so much 'such or such' an institution of power, or group, or elite, or class, but rather a technique, a form of power. . . . We have to promote new forms of subjectivity [Foucault continues] through the refusal of this kind of individuality [modern, Western, industrial, bourgeois] which has been imposed on us for several centuries."[14] But, then, on Foucault's own argument, *every* successful "resistance" will merely generate a new "subjectivity" inviting the same sort of resistance without *any* benefit from having compared the merit of competing "forms of power." Resistance, for Foucault, has *nothing* to do with the substantive forms of life that happen to be normalized within any given forms of power but only with resisting mere normalization as such. (This shows at least that Foucault's view is diametrically opposed to Rorty's.[15]) Even more disastrously, Foucault's instruction *altogether lacks specific political or moral force*, because the resistance it invites it detaches from the *critique* of the actual structures of a particular society as well as from any particular improvements recommended for that society, and because it reattaches "loyalty" only to the *myth* of sustaining the inchoate flux beneath all normalizing *epistemes*. Foucault's recommendation is active and "practical" all right, but it is neither prudential nor political nor moral in any recognizable sense. It has no ideology or program of reform. In fact, it is remarkably like Martin Heidegger's well-known conception of practical life advocated in the "Letter on Humanism"—the spirit of which, of course, Foucault would unconditionally oppose. But we must remember that Heidegger also lacked a political and moral sense in spite of his insistence on a form of supreme practical activism: "Thinking [says Heidegger] is not merely *l'engagement dans l'action* for and by beings, in the sense of the actuality of the present situation. Thinking is *l'engagement* by and for the truth of Being. *That*, Heidegger believed, *is* "the essence of action," the *Gelassenheit* in which "Being comes to language."[16] His is a "practical" commitment that has, and can have, absolutely no political or moral relevance in a comparative, distributed, competitive, engaged, praxically focused, ideologically serious, or technologically viable sense; for every would-be engagement risks, for Heidegger, "the triumph of the manipulable arrangement of a scientific-technological world and of the social order proper to that world." "Questions," Heidegger says, "are paths toward an answer. If the answer could be given it would consist in a transformation of thinking, not in a propositional statement about a matter at stake. [It would evoke our loyalty to a] thinking which can be neither metaphysics nor science [nor politics nor morality, we may add]."[17] It would commit us to what, in a sense, is forever "absent" (in Heidegger's sense): to the "opening," *aletheia*, to that which has no determinate consequences of a political or moral sort beyond that remarkable commitment itself.

Some have construed Heidegger's practical lesson as the lesson of "humility" (perhaps with respect to any and every determinate moral or political program).[18] Certainly, the theme *may* be an ingredient (even a welcome ingredient) *in* a political or moral or prudential program; but, in itself, it is essentially quietist and supervenient, conceptually dependent on and effectively fruitful only with respect to *some* antecedently determinate program, incapable in itself of supporting the kind of activism we know Foucault and Heidegger (in entirely different ways) actually espoused. The irony remains that, though both Heidegger and Foucault manifest a keen grasp of the effects of a historicized technology on values, neither is quite able to fuse a humanly pertinent morality with an understanding of how its continued relevance depends on the technologized world in which it arises and has application.

Rorty's postmodernism is also morally and politically active, but it is also inherently aimless, apologetic, convincing only in a tame and conservative and comfortable (bourgeois liberal) setting. Foucault's postmodernism is certainly active (really: even more active), but the irony remains that it is aimless in an even deeper moral or political sense—in the sense in which *it can have no determinate program of its own*, except accidentally, reactively, parasitically.[19]

The lesson is clear. Political, legal, moral programs must be historically and praxically engaged. They must address the categories and what those categories designate within the forms of life of particular societies. There *is* no political or moral program that ignores or repudiates legitimation. There can be no postmodernist politics or morality: because (1) every such program must be at least congruent with the historical possibilities of particular societies (which, of course, we may not adequately perceive), and no such program can be normatively serious if it merely confirms, actively or by default, the mere ongoing practices of such societies; and because (2) no postmodernist is consistently able (or, of course, willing) to admit that, under whatever constraints may be named—for instance, those (earlier conceded) regarding intransparency, preformation, artifactuality—the conceptual distinctions in place in any society *can* be legitimated in descriptive, explanatory, normative, or prescriptive ways. This is why, in spite of the greater flexibility of Foucault's schema when compared with Marx's, Foucault's vision is ultimately *not* a moral or political vision in the way Marx's is. It cannot be, because *it is not praxically and technologically engaged in an existential way with the perceived structure of (and the perceived alterability and replaceability of those structures within) the life of an actual society*. For related reasons, there can be no "ineffabilist" politics or morality, in the sense in which Foucault and Heidegger mean to persuade us to bring our "practical" life into some essential and overriding accord with the sheer flux of *pouvoir/savoir*, or the inexhaustibility of *Sein* before addressing any particular alethic "clearing."[20] On the admission of intransparency and the other constraints linked with it, it is certainly true that there could hardly be a uniquely valid political or moral program, even if we recovered the notion of objectivity and legitimation. But now, that admission itself becomes an instructive clue regarding what might reasonably be proposed as a condition of objectivity in political and moral commitments.

III

All that we have laid out so far may be collected in a sentence: *moral, legal, and political concerns do not arise except in the context of actual belief, and they cannot be resolved except under constraints that cognitionally affect the formation and testing of belief.* The relevance of Heidegger and Foucault, therefore, is doubly doubtful or, on the most sanguine reading, reduced to an *adverbial* qualification of whatever, on independent grounds, *would* be morally or practically substantive in the way of belief. For either: (a) the use of the notions of *aletheia* and *pouvoir/savoir* afford a mythic context for *any* pertinent beliefs that Heidegger and Foucault cannot themselves responsibly generate—hence, that their vision can do no more than qualify in an adverbial way independently legitimated action and commitment; or (b) the use of those notions entails a morality or politics of undifferentiated quietism (Heidegger) or of undifferentiated anarchism (Foucault). Furthermore, political, legal, and moral beliefs, formed under those conditions (mentioned earlier on) in which theoretical and practical reason cannot be disjoined in principle, will be either postmodern or modernist in nature or else will have to pursue an entirely novel form of legitimation—the very need for which has hardly been perceived in the usual disputes, and the form of which remains largely unknown. For: either (c) such beliefs will not require or will not be thought capable of legitimative support at all (postmodernism), however much they "prudentially" or "reasonably" continue to conform with the practices of an enveloping society (as in Rorty); or (d) the mere historical context of such beliefs will be taken to permit the fully legitimated recovery of the norms by which such beliefs are rightly ordered and appraised (modernism), however doctrinaire the values relevant efforts may try to vindicate (as in Alasdair MacIntyre, Richard Bernstein, Charles Taylor, Hans-Georg Gadamer, and Jürgen Habermas).[21]

Current views of intransparency and technological preformation oblige us to construe the quest for knowledge as a quest entirely *internal to the space of praxically formed belief.* So seen, the nostalgia for certitude that Heidegger and Foucault so energetically resist *does* still (extravagantly) affect their *own* alienation from the context of practical life. Otherwise, they would never have blundered into supposing that the seriousness with which human beings commit themselves responsibly would ever, for that reason alone, be taken to yield compelling evidence that they had succumbed to the charms of mere "presence" (in Heidegger's sense) or to the processes of "normalization" (in Foucault's).[22] Again, the same sort of nostalgia need not have been so fiercely tethered (as by Rorty) to disallow even the conceptual *need* to try to legitimate what (despite Rorty's gymnastic and boyish charm) could never be more than a lucky form of life—that is, the American or Western European—that by its well-stocked shelves might lull us (as it plainly has Rorty) into applying its own laissez-faire mentality globally. (Think of applying it, for instance, to Bangladesh or Ethiopia.) The pretense is preposterous, utterly obscene in fact.

The rational construction of a political morality is a difficult undertaking, perhaps an impossible one. But it would make no sense if it disallowed second-order legitimative work; and it would have no point if it were not effectively congruent with the real options of an actual society. It would remain blind by the first denial and empty by the second impertinence. If we acknowledge as well the triple constraints already drawn by opposing the disjunction of theoretical and practical reason—intransparency, the constructive or artifactual nature of world and self, the preformative nature of human histories—then it must be clear that the canonical forms of moral theory are very nearly bankrupt, that is, *all* versions of eudaimonism, essentialism, natural law theories, universalism, naturalism, normative realism, traditionalism, formalism, teleologized histories, "patterned" principles.[23] (What they value need not be altogether pointless, however.)

The world is a changing motley; and reason and human nature are technologized or praxical variants within its *epistemes.* Thus far, the postmodernists are entirely right. Nevertheless, the "indicative" *minima* of moral reflection do remain: for example, the general biologized concerns of prudence, contingent confinement to the planet, the remembered histories and aspirations of different peoples, the actual use and distribution of the earth's resources, the necessity of socially shared practices and ideologies.[24] These *minima*—detailed, distributed, local, variable, contingent—yield no more than *indicative universals* (hence no universalisms), historically restricted or nearly vacuously generic invariances that remain pertinent to practical life but entail no normative legitimation in themselves: for instance, the variable life expectancies of different societies of mortal humans (*all* mortal), or the different convictions of different peoples regarding condemning or accepting suicide or homosexuality or private property.

In the sciences, it is the radical *under*determination of causal and explanatory theories that, on any argument, commits us rationally to second-order legitimative bets.[25] In political morality, it is the *over*determination of competitive moral convictions (often opposed to prudence) that induces us to seek to form a legitimative rationale for policies deliberately undertaken. That is the theme that runs from Hobbes to Marx—that redeems the modernist conviction (against the postmodernist), that confirms the inescapability of second-order argument. It is only the outmoded zeal of the modernist that must be curbed—the vestigial conviction, for instance, of traditionalists like Gadamer (who holds classical Greek humanity to be normatively invariant for the whole of Asia and Africa) or of late-Enlightenment sports like Habermas and Karl-Otto Apel (who hold that, asymptotically, emancipatory reason progressively approaches the fixed norms of universal moral intelligence).[26] The absurdity and arbitrariness of these options are plain enough.

The important point to grasp here is that what is being exposed is the profound lacuna of the whole of Western moral theorizing: the incantatory, almost self-congratulatory refusal to come to terms with the meaning of a praxical conception of life, the significance of a technologized history. We must bring our legitimative theories into line with the saliencies (the indicative invariances) of our age—they need not be convergent; and we must learn to live with the deep fact of irreconcilable policies and irreconcilable legitimative programs. Hegel remarks, in the *Philosophy of Fine Art*, that the cosmic order may, as in Sophocles' *Antigone*, indifferently destroy the human agents, if it must, in order to restore its rightful harmony.[27] He is right, of course, except that there *is* no higher cosmic order that must or will be restored, there *are* no agents but human agents, history *has* no purpose or direction of its own, there *are* no uniquely or ideally adequate resolutions of human conflict, there *are* no invariant principles in terms of which the gains and losses of any action can be objectively assessed, there *is* no higher reason, there *is* no exiting from history.

In fact, the myths and literature of the world, if they have anything to teach at all, confirm, within the generic uniformities of human life, the sheer variety of viable human practice and compelling ideal (or ideology). *Any* move to extract uniquely adequate, univocal, exceptionless second-order *norms* from that first-order variety is an empirical cheat, a plain violation of the conditions of *praxis* and technology (that have no determinate legitimative force of their own).[28]

The usual errors are these: first, the conflating of, and the confusion between, generic prudential concerns (life, food, shelter, security, companionship, property) and the historically contingent, distributed, disputatious, alterable instantiations of those concerns; second, the failure to recognize that the historicized nature of human values embedded in actual practice neither disallows an escape *from* particular historical horizons (as by revolution, revelation, reform) nor permits an escape *to* non-historical or posthistorical or reliably transhistorical norms (as advocated, for instance, in different ways, by Leo Strauss, by Habermas, by Gadamer); and third, the utterly unguarded, arbitrary denial that legitimative arguments can be rationally defended *in* a world in which transparency is denied, in which the artifactuality of man and his world is admitted, in which second-order arguments are also incapable of escaping one historicized horizon or another.

IV

What is needed is the construction of a notion of *rationality* fitted to the constraints we have repeatedly acknowledged (intransparency, artifactuality, preformation); capable of supporting legitimative arguments; articulated for both science and political morality; completely opposed to prevailing modernist and postmodernist pretensions. *There is no such construction at the present time.* But in the same spirit in which the historicized rationales of scientific research programs are being recovered and tested (as they must, on pain of construing science as completely arbitrary or chaotic),[29] we may also marshal an informal vision of the legitimation of practical institutions.

Of course, the single most arresting event in which the question of the nature of human rationality has been forced on the attention of modern man is the French Revolution. For, as a result of the Revolution and its sequelae, all the normative or legitimative fixities of human nature have been utterly subverted, that is, the Enlightenment vision as well as the more ancient models of natural reason from which, admittedly, the Enlightenment departed in its own radical way.[30] The would-be legitimative invariances of reason—meant to serve science and morality equally, as (in different ways) in Aristotle, St. Thomas, Descartes, and Kant—have been completely subverted (by way of the slowly dawning import of the French Revolution) by the indicative, internal, historicized, interested, horizonal ideologies of different ages and shifting social classes.

The supreme lesson of the Revolution is this: *that human reason is itself a contingent but hardly arbitrary artifact internal to the historicized life of particular societies.* The supreme lesson of our pondering that lesson is this: *that reason can and must, nevertheless, perform a second-order legitimative function.* "Reason" is both an artifact and a norm of historical life. That is the double lesson we have not yet learned. It is contested, of course, in Claude Lévi-Strauss' hopeless quarrel with Jean-Paul Sartre about the impossibility of a historicized human science.[31] It is contested in Habermas's endless machinations designed to extricate human reason from the snares of dialectical history that Habermas's own reading of Marx and the Frankfurt School theorists should have strictly disallowed.[32] The worry is clear enough, but the corrective—the modernist corrective—is quite untenable.

The essential clue to its failure is inadvertently given in the following strenuous (modernist) comment by Hilary Putnam, intended to escape all forms of relativism:

> Philosophers who lose sight of the immanence of reason, of the fact that reason is always relative to context and institution, become lost in characteristic philosophical fantasies. . . . Philosophers who lose sight of the transparency of reason become cultural (or historical) relativists.[33]

There is no essential difference between Putnam's argument and the arguments of Lévi-Strauss and Habermas. Putnam's argument fails for an elementary reason (as, correspondingly, do theirs): the "regulative" function of reason (that is, the critical or legitimative or pragmatic or transcendental function) may, *qua* regulative, escape *some* putative historical horizon h_1; but, in doing that, *on* Putnam's own entirely reasonable sketch of the "immanence" of reason, it cannot possibly escape *another* historical horizon h_2. The important point to understand is that *it need not escape* to be regulative in the second-order sense; and that, in that case, legitimative discourse (the regulative function of reason) cannot preclude relativism, perspectivism, historicism, internalism, intransparency.[34] It can be legitimative but it cannot be transcendental in the Kantian sense—synthetic *a priori*. It is only the nostalgia for the universalisms of transparency that have kept us from grasping the fact that the entire history of legitimative inquiry can be recovered once the pretensions of objectivisms, logocentrisms, privileged access, and the like have been repudiated. There is the clue to the postmodernist's mistake.

V

We have been speculating under the umbrella of the French Revolution. We have recovered the coherence of political morality under its terms; but we have found its terms to be quite strenuous. We must not suppose that the indicative invariances of *Homo sapiens* entail any second-order norms fitted to the methods of science or the practices of societal life.

Prudence is biologically grounded and culturally determined. On its biological side, it is never more than generic. For example, the sheer prolongation of life is not strictly invariant within the human species, contrary to Baruch Spinoza's or Sigmund Freud's *conatus*—hence, also, *not* determinative of essential human reason. It is not even demonstrable, for instance, that suicide cannot constitute a rational decision in the face of the conditions of actual mortal life. It cannot be merely a disorder of reason.[35]

Distributed constraints on suicide are invariably culture-specific, also remarkably different among different societies—as the difference between Stoic and Christian imagination shows. The same is true for *every* would-be prudential norm addressed to our admitted biological concerns. There, for instance, is the fatal weakness of the natural rights theory.[36] Persons (unlike *Homo sapiens*) do not form natural kinds: speaking apes, Martians, vampires, suitably advanced machines may all be persons. They must be physically embodied in some way or another, but they need not be confined to a single natural species and they need not even be confined to natural biological kinds.[37] That persons have no *natures* (though they are inseparably manifested in physical bodies that normally belong to natural kinds—*Homo sapiens*), that they have only *histories*, certainly complicates the legitimation of political moralities; *but it does not preclude it.*[38]

To admit the conceptual failing of the natural rights doctrine—which is not the same thing as admitting its political or legal or moral failing—does indeed raise and answer the same question that Hobbes's notion of the "laws of nature" is meant to raise and answer. For: *there are no such laws or norms*. The artifactuality of human selves is against it; although that same artifactuality could still support a prudential realism—in at least two respects: first, because no politically pertinent convictions could possibly stray very far from the generic conditions of the viability of the species (the conditions of prudence)—even among such marginal societies as those of the Shakers and the Thugs and the Ik; secondly, because the realism of human culture (unlike the realism of physical nature) is essentially what and only what, critically and consensually, *appears* to be the forms of life of given societies.[39] The minimal realism of political morality is simply the realism of human culture itself—the common theme that draws from such different visions as the naturalism of Ludwig Wittgenstein's *Lebensformen* and the phenomenology of Hegel's *Sitten*. Societies are at least what they appear to be to their own apt members. Their presumption of true virtues, true duties, true rights, true forms of self-realization is another matter altogether. The realism of the first sort, one's form of life, is simply the realism of historicized prudence; the realism of the second, an actual society's *telos* or rationality, requires a legitimative leap *that the first cannot possibly entail, and that nothing can secure.*

Moral and political disputes are dialectically constrained by generic prudence and the reflexive saliencies of actual cultural life. They never yield more than the indicative invariances of a particular horizon. Hobbes solves the implied problem by falling back to a science of the invariances of prudence—to a simple objectivism of norms; and Marx solves the problem by falling back to a teleologized history of social development—to a privilege of prophecy. Hobbes, after all, is a premodernist, and Marx a modernist; and both fail to grasp the peculiar logic of practical reason. Political morality cannot *discover* any canons. The artifactual nature of cultural existence precludes any finding of *universal law or rule or principle or norm of prudence*. It also precludes the discovery of *any essential telos or morality or rational practice*. There are none to be discovered.

We say this casually enough, but it is a terrible truth. The meaning of life cannot be found except *in* one's horizoned world; but it cannot be *found* there. The rule is that there *is* no rule. Practical reason is dialectical, marked by what it cannot claim. It cannot claim a direct ethical or legal or political realism of rules or norms. It cannot claim a human essence. It cannot claim prudential *minima* except generically and variably and conditionally and even incompatibly. It cannot claim validity for whatever cannot attract the sizable and ideological energies of a people, but it cannot be legitimated by that alone. It cannot preclude incompatible norms or incompatible legitimative rationales. It cannot claim an ideal convergence or a universal consensus. It cannot preclude relativism. It cannot claim adequate rules for adjudicating among opposing judgments, commitments, policies, programs. It cannot confirm an exclusive individualism or collectivism, or fix the right priorities within or between individual and collective goods. It cannot claim moral neutrality.

The plausibility of these charges is tied to our post-Revolutionary mentality. The entire project of political morality (of the whole of practical reason) is governed by the loss of transparency, a constructed world and self, the historical preformation of reflexive life. There can be no divide between science and politics, once the disjunction of theory and practice is denied. Human nature is its own open-ended history; and its history includes its own self-transforming revolutions. The order of history is the history of reconstituting order again and again, divergently and incompatibly, within the constraints of local coherence, and of reconstituting again and again its would-be history. That is the lesson of the Revolution. It is an order of construction, not of discovery. Practical reason is a name for the endless forms of devotion to the continuance of human life under the perception of its viability and supposed direction.

Temple University

NOTES

1. Thomas Hobbes, *Leviathan*, ed. Michael Oakeshott (Oxford: Basil Blackwell, n.d.), chapter 5, pp. 28, 30.
2. The first is developed, for instance, in different ways, in Thomas S. Kuhn, *The Structure of Scientific Revolutions*, 2d ed. (Chicago: University of Chicago Press, 1970), and in Nelson Goodman, *Languages of Art* (Indianapolis: Bobbs-Merrill, 1968). The second is developed, again in different ways, in Nancy Cartwright, *How the Laws of Physics Lie* (Oxford: Clarendon, 1983), and in Bas C. van Fraassen, *The Scientific Image* (Oxford: Clarendon, 1980).
3. See Joseph Margolis, "Berkeley and Others on the Problem of Universals," in Colin M. Turbayne, ed., *Berkeley: Critical and Interpretive Essays* (Minneapolis: University of Minnesota Press, 1982), and *Pragmatism without Foundations: Reconciling Realism and Relativism* (Oxford: Blackwell, 1986), chapter 11.
4. For a fuller account of the role of *praxis*, see Joseph Margolis, *Texts without Referents: Reconciling Science and Narrative* (Oxford: Blackwell, 1988), chapter 4.
5. Michel Foucault, *The Order of Things* (New York: Vintage, 1973), pp. 308-309.
6. See Nicholas Lobkowicz, *Theory and Practice: History of a Concept from Aristotle to Marx* (Notre Dame: University of Notre Dame Press, 1967).
7. Roberto Mangabeira Unger, *Social Theory: Its Situation and Its Task; A Critical Introduction to Politics, a Work in Constructive Social Theory* (Cambridge: Cambridge University Press, 1987), p. 191; see the rest of the chapter.
8. This, of course, counts against the arguments, incomplete in any case, of Richard Rorty, *Philosophy and the Mirror of Nature* (Princeton: Princeton University Press, 1979); and Jean-Francois Lyotard, *The Postmodern Condition: A Report on Knowledge*, trans. Geoff Bennington and Brian Massumi (Minneapolis: University of Minnesota Press, 1984).
9. See Joseph Margolis, *Texts without Referents*; chapter 6.
10. See Joseph Margolis, *Science without Unity: Reconciling the Human and Natural Sciences* (Oxford: Blackwell, 1987).
11. I gave a brief overview of postmodernist strategies in this regard, in "A Convergence of Pragmatisms," presented at the conference: "Frontiers in American Philosophy," Texas A & M University, College Station, Texas, Spring 1988.
12. Richard Rorty, "Postmodernist Bourgeois Liberalism," *Journal of Philosophy* 80 (1983); reprinted in Robert Hollinger, ed., *Hermeneutics and Praxis* (Notre Dame: University of Notre Dame Press, 1985), pp. 216-217.

13. Michel Foucault, "The Subject and Power," afterword to Hubert L. Dreyfus and Paul Rabinow, *Michel Foucault: Beyond Structuralism and Hermeneutics* (Chicago: University of Chicago Press, 1982), p. 208. The first part of the essay was written in English; the second part was translated by Leslie Sawyer. See, also, Luther H. Martin, Huck Gutman, and Patrick H. Hutton, eds., *Technologies of the Self: A Seminar with Michel Foucault* (Amherst: University of Massachusetts, 1988).

14. Foucault, "The Subject and Power," pp. 212, 216.

15. It is also, of course, remotely related to the issue raised in Herbert Marcuse, *One-Dimensional Man: Studies in the Ideology of Advanced Industrial Society* (Boston: Beacon Press, 1964), which implicitly opposes Rorty's thesis as well.

16. Martin Heidegger, "Letter on Humanism," trans. Frank A. Capuzzi, with J. Glenn Gray, in *Basic Writings*, ed. David Farrell Krell (New York: Harper and Row, 1977), pp. 193-194. See, also, "The End of Philosophy and the Task of Thinking," trans. Joan Stambaugh, in the same volume. See also Otto Poggeler, "Heideggers politisches Selbstverstandnis," in Annemarie Gethmann-Siefert und Otto Poggeler, eds., *Heidegger und die praktische Philosophie* (Frankfurt am Main: Suhrkamp, 1988).

17. Heidegger, "The End of Philosophy and the Task of Thinking," pp. 373, 377, 378.

18. *Ibid.*, pp. 386-387. See John D. Caputo, *Radical Hermeneutics: Repetition, Deconstruction, and the Hermeneutic Project* (Bloomington: Indiana University Press, 1987), p. 258. The point is not meant in a necessarily ethical sense.

19. For postmodernist opposition to laissez-faire political morality, one must turn to Paul Feyerabend, not to Foucault; but there, the radical *optimism* expressed is entirely anarchical. *Every* political or moral benefit it affords is entirely gratuitous, open to appraisal on grounds that it cannot and would never, in principle, supply. See Paul Feyerabend, *Science in a Free Society* (London: NLB, 1978); and *Farewell to Reason* (London: Verso, 1987).

20. I offer a fuller analysis of ineffabilism in "The Critical and Mythic Strategies of Intransparency," delivered at the conference: "Beyond Translation," University of Warwick, Warwick, England, Summer 1988.

21. See, for instance, Richard J. Bernstein, *Beyond Objectivism and Relativism: Science, Hermeneutics, and Praxis* (Philadelphia: University of Pennsylvania Press, 1983); and Charles Taylor, "Philosophy and Its History," in Richard Rorty, J. B. Schneewind, and Quentin Skinner, eds., *Philosophy in History: Essays on the Historiography of Philosophy* (Cambridge: Cambridge University Press, 1984).

22. See Martin Heidegger, *Being and Time*, trans. John Macquarrie and Edward Robinson (New York: Harper and Row, 1962), pp. 25-26; and Michel Foucault, "Panopticism," *Discipline and Punish: The Birth of the Prison*, trans. Alan Sheridan (New York: Vintage Books, 1979).

23. On patterned principles, see Robert Nozick, *Anarchy, State, and Utopia* (New York: Basic Books, 1974), p. 156.

24. See Joseph Margolis, *Negativities; The Limits of Life* (Columbus, Ohio: Merrill, 1975).

25. One may have a conveniently quick sense of the extent to which such underdetermination obtains in current physics from Stephen W. Hawking, *A Brief History of Time: from the Big Bang to Black Holes* (New York: Bantam Books, 1987), particularly chapter 8.

26. See Hans Georg Gadamer, *Truth and Method*, trans. Garrett Barden and John Cumming (New York: Seabury Press, 1975), second part, II, ii; and Jürgen Habermas, *Reason and the Rationalization of Society*, trans. Thomas McCarthy (Boston: Beacon Press, 1984), chapter 1.

27. See G. W. F. Hegel, *Philosophy of Fine Arts*, 4 vols., trans. F. P. B. Omaston (London: Bell, 1920).

28. See, for an opposing view, Leo Strauss, *Natural Right and History* (Chicago: University of Chicago Press, 1953); Leo Strauss and Joseph Cropsey, eds., *History of Political Philosophy*, 3d ed. (Chicago: University of Chicago Press, 1987), epilogue; and Martha Nussbaum, *The Fragility of Goodness: Luck and Ethics in Greek Tragedy and Philosophy* (Cambridge: Cambridge University Press, 1985).

29. See, for instance, Ian Hacking, *Representing and Intervening: Introductory Topics in the Philosophy of Natural Science* (Cambridge: Cambridge University Press, 1983); Richard N. Boyd, "Realism, Approximate Truth and Philosophical Method," in C. Wade Savage, ed., *Scientific Theories* (Minnesota Studies in the Philosophy of Science, vol. 14; Minneapolis: University of Minnesota Press, 1989).

30. See Alasdair MacIntyre, *After Virtue*, 2d ed. (Notre Dame: University of Notre Dame Press, 1984).

31. See Claude Lévi-Strauss, *The Savage Mind* (Chicago: University of Chicago Press, 1966), chapter 9.

32. See Jürgen Habermas, *The Philosophical Discourse of Modernity: Twelve Lectures*, trans. Frederick Lawrence (Cambridge: MIT Press, 1987); and Max Horkheimer and Theodor W. Adorno, *Dialectic of Enlightenment*, trans. John Cumming (New York: Seabury Press, 1972).

33. Hilary Putnam, "Why Reason Can't Be Naturalized," *Synthese* 52 (1952); reprinted in Kenneth Baynes *et al.*, eds., *After Philosophy: End or Transformation?* (Cambridge: MIT Press, 1987), p. 228.

34. On transcendental arguments, see Margolis, *Pragmatism without Foundations*, chapter 11.

35. See *The Standard Edition of the Complete Psychological Works of Sigmund Freud*, ed. James Strachey *et al.* (London: Hogarth Press and the Institute of Psycho-analysis, 1966-1974), *An Outline of Psycho-Analysis*, chapter 2, vol. 23; Freud's letter to Albert Einstein, "Why War?"; also, Edwin S. Shneidman, Norman L. Farberow, and Robert E. Litman, *The Psychology of Suicide*, chapter 37; and Margolis, *Negativities*, chapter 2.

36. See Margolis, *Negativities*, chapter 10.

37. See Joseph Margolis, *Persons and Minds: The Prospects of Nonreductive Materialism* (Dordrecht: Reidel, 1978), chapter 7.

NAME INDEX

This index collects only the names in the text. The notes the authors have appended to each contribution are an excellent source for additional references.

Aas, Gro Hanne 11
Akrich, Madeleine 55, 57
Amy, Douglas 116, 118
Anderson, Richard 112
Apel, Karl-Otto 226
Aquinas, Thomas 227
Aristotle 189, 201, 216-217
Aubenque, Pierre 194

Bacon, Francis 16, 164
Barber, Benjamin 108-109
Barthes, Roland 56, 59
Baudrillard, Jean 59
Bedny, Demyan 181
Beiner, Ronald 109
Bellah, Robert 210-211
Bellamy, Edward 91
Bentham, Jeremy 127
Berger, Peter 83-84
Bernstein, Richard 225
Bijker, Wiebe 54-55, 57, 60
Billington, David 208
Borgmann, Albert 9, 81
Bradley, F. H. 164-165
Braudel, Fernand 56
Brenna, Brita 11
Breton, Jean-Louis 60
Brock, Dan W. 207
Bryce, Lord James 162
Brzezinski, Zbigniew 96
Bukharin, N. I. 179
Bullert, Gary 85

Carnap, Rudolf 108-109
Carpenter, Stanley R. 8
Carter, Jimmy 96
Castoriadis, Cornelius 38, 40
Cato 46
Cérézuelle, Daniel 11
Chandler, Alfred D. 53
Charbonneau, Bernard 45
Clifford, W. K. 132
Colglazier, William 111
Comte, Auguste 165
Cowan, Ruth Schwartz 54
Crumrine, Tod 113

Davis, Gary 111
De Tocqueville, Alexis 99, 161-163, 169
DeGaulle, Charles 63, 68-69
Della-Voss, Viktor K. 176
Descartes, René 164
Dewey, John 7, 20, 22, 81, 85-86, 91-95, 97-101, 164
Dickson, David 168
Dubreuil, Hyacinthe 60
Durbin, Paul T. 7
Durkheim, Emile 165
Duverger, M. 39
Dworkin, Ronald 209

Einstein, Albert 28, 177
Eisenhower, Dwight D. 16, 97
Elliott, Michael 115
Ellul, Jacques 6-7, 21, 88, 105-109, 208
Emerson, Ralph Waldo 91
Engelmeier, Pyotr K. 176-178, 184

Farias, Victor 1087
Feuerbach, Ludwig 32
Fielder, John 7-8
Forbes, H. D. 109
Ford, Henry 61
Foucault, Michel 10, 56, 217, 221-225
Frege, Gottlob 164
Freud, Sigmund 32, 229
Frost, Robert L. 6-7

Gadamer, Hans-Georg 225-227
Galbraith, John Kenneth 91, 97
Genovese, Eugene 83
Godard, Jean-Luc 70
Goethe, J. W. 195
Gore, Albert 33
Gorokhov, Vitaly 9
Green, T. H. 165
Greenberg, Michael 112
Gulbrandsen, Elisabeth 11

Habermas, Jürgen 225-228

235

NAME INDEX

Hansen, James 33
Hardin, Garrett 162
Harsanyi, John 124-135
Hegel, G. W. F. 32, 165, 189, 192-193, 198, 226, 230
Heidegger, Martin 9, 164, 168, 187-203, 222-225
Herder, J. G. 164
Hickman, Larry A. 7, 11
Hitler, Adolf 9, 30
Hobbes, Thomas 164-165, 215-220, 226, 229-230
Høegh-Omdal, Kathrine 11
Holderlin, Friedrich 198
Hoover, Herbert 92
Hughes, Thomas P. 53-54, 66
Humboldt, K. W. 164
Hume, David 127
Hunt, Lynn 59
Husserl, Edmund 192

Imperatore, Mary 11

James, William 31, 93, 100, 196
Johnson, Michael 49
Jonas, Hans 209

Kahneman, Daniel 134
Kant, Immanuel 8, 127, 143, 189, 220, 227
Kasperson, Roger 111
Khudyakov, P. K. 177
Kierkegaard, Soren 201
Kolakowski, Leszek 167
Korolev, Sergei 182

Lands, David 53
Latour, Bruno 55
Law, John 53
Lenin, V. I. 177
Lévi-Strauss, Claude 56, 228
Levinger, David 11
Locke, John 127, 164-165
Lovekin, David 49
Luckmann, Thomas 83-84
Lukács, Georg 187
Lukes, Stephen 165

MacIntyre, Alasdair 210-211, 225
Marcuse, Herbert 84, 86
Margolis, Joseph 10, 168
Marx, Karl 10, 21, 31-32, 81-82, 88, 164-168, 217, 223, 226, 228,

Massé, Pierre 69
McDermott, John 7, 91, 94-101
Mead, George Herbert 7, 32, 81, 86, 165
Mesthene, Emmanuel 7, 91, 94-99, 101-102
Michalos, Alex 132
Mill, John Stuart 127
Monnet, Jean 63
Moser, Ingunn 11
Moynihan, Daniel P. 99

Nelkin, Dorothy 112
Nielsen, Torben Hviid 11
Noble, David 54, 84, 88

Orwell, George 98
Otway, Harry 113

Pacey, Arnold 52
Partant, Francois 45
Paul, Marcel 63, 68
Pfaffenberger, Bryan 56
Plato 46, 99, 177, 195, 201
Polanyi, Karl 167
Pollak, Michael 112
Putnam, Hilary 228

Rawls, John 125-128, 130, 209
Rockmore, Tom 9
Roosevelt, Franklin D. 28
Rorty, Richard 10, 221-225
Ross, Gail 11
Rousseau, Jean-Jacques 8, 127, 143, 164, 166
Russell, Bertrand 164
Rykov, A. I. 177

Sartre, Jean-Paul 192, 201, 228
Schrank, Robert 146
Sclove, Richard E. 8
Sejersted, Francis 11
Sen, Amartya 127, 131
Shrader-Frechette, Kristin 8
Skolimowski, Henryk 208
Smart, J. J. C. 127
Smith, Adam 127
Socrates 30, 32
Sophocles 226
Spinoza, Baruch 229
Stalin, Joseph 9, 63, 178-179, 182
Strauss, Leo 227

Taylor, Charles 162, 164-165, 225
Thurow, Lester 209
Tichi, Cecelia 91-92
Toffler, Alvin 169
Trotsky, Leon 85
Tupolev 182
Turkle, Sherry 59, 141
Tversky, Amos 134

Unger, Roberto M. 218

Veblen, Thorstein 3, 65, 91
Vico, Giambattista 32

Wartofsky, Marx 5, 7
Watkin, J. W. N. 133-134
Weber, Max 191
Wiener, Norbert 16
Winner, Langdon 54, 105-106, 108-109, 118, 167
Wittgenstein, Ludwig 164, 230

PHILOSOPHY AND TECHNOLOGY
Series Editor: Paul T. Durbin

OFFICIAL PUBLICATIONS OF
THE SOCIETY FOR PHILOSOPHY AND TECHNOLOGY

1. *Philosophy and Technology*
   Edited by Paul T. Durbin and Friedrich Rapp. 1983   ISBN 90-277-1576-9
   (Published as Volume 80 in 'Boston Studies in the Philosophy of Science')
2. *Philosophy and Technology, II.* Information Technology and Computors in Theory and Practice.
   Edited by Carl Mitcham and Alois Huning. 1986   ISBN 90-277-1975-6
   (Published as Volume 90 in 'Boston Studies in the Philosophy of Science')
3. *Technology and Responsibility*
   Edited by Paul T. Durbin. 1987   ISBN 90-277-2415-6; Pb 90-277-2416-4
4. *Technology and Contemporary Life*
   Edited by Paul T. Durbin. 1988   ISBN 90-277-2570-5; Pb 90-277-2571-3
5. *Technological Transformation.* Contextual and Conceptual Implications
   Edited by Edmund F. Byrne and Joseph C. Pitt. 1989   ISBN 90-277-2826-7
6. *Philosophy of Technology.* Practical, Historical and Other Dimensions
   Edited by Paul T. Durbin. 1989   ISBN 0-7923-0139-0
7. *Broad and Narrow Interpretations of Philosophy of Technology*
   Edited by Paul T. Durbin. 1990   ISBN 0-7923-0684-8
8. *Europe, America, and Technology: Philosophical Perspectives*
   Edited by Paul T. Durbin. 1991   ISBN 0-7923-1254-6
9. *Democracy in a Technological Society*
   Edited by Langdon Winner. 1992   ISBN 0-7923-1995-8

KLUWER ACADEMIC PUBLISHERS – DORDRECHT / BOSTON / LONDON